Isabel Hornibrook

Camp and trail - A story of the Maine woods

Isabel Hornibrook

**Camp and trail - A story of the Maine woods**

ISBN/EAN: 9783743338258

Manufactured in Europe, USA, Canada, Australia, Japa

Cover: Foto ©berggeist007 / pixelio.de

Manufactured and distributed by brebook publishing software (www.brebook.com)

Isabel Hornibrook

**Camp and trail - A story of the Maine woods**

THE MOOSE WAS NOW SNORTING LIKE A WAR-HORSE BENEATH.

(*See page* 274.)

# A Story of the Maine Woods

BY

## ISABEL HORNIBROOK

AUTHOR OF "TUKE," "IN THE SERVICE," "LOST IN MAINE WOODS," ETC.

---

BOSTON
LOTHROP PUBLISHING COMPANY

TO
J. L. H.

# PREFACE.

IN adding another to the list of stories bearing on that subject of perennial interest to boys, adventures in camp and on trail among the woods and lakes of Northern Maine, one thought has been the inspiration that led me on.

It is this : To prove to high-mettled lads, American, and English as well, that forest quarters, to be the most jovial quarters on earth, need not be made a shambles. Sensation may reach its finest pitch, excitement be an unfailing fillip, and fun the leaven which leavens the camping-trip from start to finish, even though the triumph of killing for triumph's sake be left out of the play-bill.

"There is a higher sport in preservation than in destruction," says a veteran hunter, whose forest experiences and descriptions have in part enriched this story. I commend the opinion to boy-readers, trusting that they may become "queer specimen sportsmen," after the pattern of Cyrus Garst ; and find a more entrancing excitement in studying the live wild things of the forest than in gloating over a dying tremor, or examining a senseless mass of horn, hide, and hoofs, after the life-spring which worked the mechanism has been stilled forever.

One other desire has trodden on the heels of the first: That Young England and Young America may be inspired with a wish to understand each other better, to take each other frankly and simply for the manhood in each; and that thus misconception and prejudice may disappear like mists of an old-day dream.

ISABEL HORNIBROOK.

# CONTENTS.

| CHAPTER | | PAGE |
|---|---|---|
| I. | Jacking for Deer | 9 |
| II. | A Spill-out | 20 |
| III. | Life in a Bark Hut | 27 |
| IV. | Whither Bound? | 41 |
| V. | A Coon Hunt | 57 |
| VI. | After Black Ducks | 72 |
| VII. | A Forest Guide-Post | 92 |
| VIII. | Another Camp | 101 |
| IX. | A Sunday among the Pines | 124 |
| X. | Forward All! | 132 |
| XI. | Beaver Works | 145 |
| XII. | "Go it, Old Bruin!" | 157 |
| XIII. | "The Skin is Yours" | 172 |
| XIV. | A Lucky Hunter | 181 |
| XV. | A Fallen King | 196 |
| XVI. | Moose-Calling | 215 |
| XVII. | Herb's Yarns | 231 |
| XVIII. | To Lonelier Wilds | 247 |
| XIX. | Treed by a Moose | 257 |

| CHAPTER | | PAGE |
|---|---|---|
| XX. | Dol's Triumph | 283 |
| XXI. | On Katahdin | 291 |
| XXII. | The Old Home-Camp | 304 |
| XXIII. | Brothers' Work | 318 |
| XXIV. | "Keeping Things Even" | 326 |
| XXV. | A Little Caribou Quarrel | 335 |
| XXVI. | Doc Again | 351 |
| XXVII. | Christmas on the Other Side | 359 |

# LIST OF ILLUSTRATIONS.

|  | PAGE |
|---|---|
| The Moose was now snorting like a War-horse Beneath | *Frontispiece* |
| "There is Moosehead Lake" | 53 |
| Dol sights a Friendly Camp | 105 |
| In the Shadow of Katahdin | 149 |
| "Go it, Old Bruin! Go it while You can!" | 169 |
| "Herb Heal" | 187 |
| A Fallen King | 199 |
| The Camp on Millinokett Lake | 239 |
| "Herb charged through the Choking Dust-clouds" | 315 |
| Greenville, — "Farewell to the Woods" | 355 |

# CAMP AND TRAIL.

## CHAPTER I.

### JACKING FOR DEER.

"NOW, Neal Farrar, you've got to be as still as the night itself, remember. If you bounce, or turn, or draw a long breath, you won't have a rag of reputation as a deer-hunter to take back to England. Sneeze once, and we're done for. That means more diet of flapjacks and pork, instead of venison steaks. And I guess your city appetite won't rally to pork much longer, even in the wilds."

Neal Farrar sighed as if there was something in that.

"But, you know, it's just when an unlucky fellow would give his life not to sneeze that he's sure to bring out a thumping big one," he said plaintively.

"Well, keep it back like a hero if your head bursts in the attempt," was the reply with a muffled laugh. "When you know that the canoe is gliding along somehow, but you can't hear a sound or feel a motion, and you begin to wonder whether you're in the air or on water, flying or floating, imagine that you're the ghost of some old Indian hunter who used to jack for deer on Squaw Pond, and be stonily silent."

"Oh! I say, stop chaffing," whispered Neal impetuously. "You're enough to make a fellow feel creepy before ever he starts. I could bear the worst racket on earth better than a dead quiet."

This dialogue was exchanged in low but excited voices between a young man of about one and twenty, and a lad who was apparently five years his junior, while they waded knee-deep in water among the long, rank grasses and circular pads of water-lilies which border the banks of Squaw Pond, a small lake in the forest region of northern Maine.

The hour was somewhere about eleven

o'clock. The night was intensely still, without a zephyr stirring among the trees, and of that wavering darkness caused by a half-clouded moon. On the black and green water close to the bank rocked a light birch-bark canoe, a ticklish craft, which a puff might overturn. The young man who had urged the necessity for silence was groping round it, fumbling with the sharp bow, in which he fixed a short pole or "jack-staff," with some object — at present no one could discern what — on top.

"There, I've got the jack rigged up!" he whispered presently. "Step in now, Neal, and I'll open it. Have you got your rifle at half-cock? That's right. Be careful. A fellow would need to have his hair parted in the middle in a birch box like this. Remember, mum's the word!"

The lad obeyed, seating himself as noiselessly as he could in the bow of the canoe, and threw his rifle on his shoulder in a convenient position for shooting, with a freedom which showed he was accustomed to firearms.

At the same time his companion stepped into the canoe, having first touched the dark object on the pole just over Neal's head. Instantly it changed into a brilliant, scintillating, silvery eye, which flashed forward a stream of white light on a line with the pointed gun, cutting the black face of the pond in twain as with a silver blade, and making the leaves on shore glisten like oxidized coins.

The effect of this sudden illumination was so sudden and beautiful that the boy for a minute or two held his rifle in unsteady hands while the canoe glided out from the bank. An exclamation began in his throat which ended in an indistinct gurgle. Remembering that he was pledged to silence, he settled himself to be as wordless and motionless as if his living body had become a statue.

From his position no revealing radiance fell on him. He sat in shadow beside that glinting eye, which was really a good-sized lantern, fitted at the back with a powerful silvered reflector, and in front with a glass lens, the light being thrown directly ahead. It was provided also with a sliding door that could be noiselessly slipped over the glass with a touch, causing the blackness of a total eclipse.

This was the deer-hunters' "jack-lamp," familiarly called by Neal's companion the "jack."

And now it may be readily guessed in what thrilling night-work these canoe-men are engaged as they skim over Squaw Pond, with no swish of paddle, nor jar of motion, nor even a noisy breath, disturbing the brooding silence through which they glide. They are "jacking" or "floating" for deer, showing the radiant eye of their silvery jack to attract any antlered buck or graceful doe which may come forth from the screen of the forest to drink at this quiet hour amid the tangled grasses and lily-pads at the pond's brink.

Now, a deer, be it buck, doe, or fawn in the spotted coat, will stand as if moonstruck, if it hears no sound, to gaze at the lantern, studying the meteor which has crossed its world as an astronomer might investigate a rare, radiant comet. So it offers a steady mark for the sportsman's bullet, if he can glide near enough to discern its outline and take aim. There is one exception to this rule. If the wary animal has ever been startled by a shot fired from under the jack, trust him never to watch a light again, though it shine like the Kohinoor.

As for Neal Farrar, this was his first attempt at playing the part of midnight hunter; and I am bound to say that — being English born and city bred — he found the situation much too mystifying for his peace of mind.

He knew that the canoe was moving, moving rapidly; for giant pines along the shore, looking solid and black as mourning pillars, shot by him as if theirs were the motion, with an effect indescribably weird. Now and again a gray pine stump, appearing, if the light struck it, twice its real size, passed like a shimmering ghost. But he felt not the slightest tremor of advance, heard no swish or ripple of paddle.

A moisture oozed from his skin, and gathered in heavy drips under the brim of his hat, as he began to wonder whether the light bark skiff was working through the water at all, or skimming in some unnatural way above it. For the life of him he could not settle this doubt. And, fearful of balking the expedition by a stir, he dared not turn his head to investigate the doings of his comrade, Cyrus Garst.

Cyrus, though also city bred, was an American, and evidently an old hand at the present business. The Maine wilds had long been his playground. He had studied the knack of noiseless paddling under the teaching of a skilled forest guide until he fairly brought it to perfection. And, in perfection, it is about the most wizard-like art practised in the nineteenth century.

The silent propulsion was managed thus: the grand master of the paddle gripped its cross handle in both hands, working it so that its broad blade cut the water first backward then forward so dexterously that not even his own practised hearing could detect a sound; nor could he any more than Neal feel a sensation of motion.

The birch-bark skiff skimmed onward as if borne on unseen pinions.

To Neal Farrar, who had been brought up amid the tumult of rival noises and the practical surroundings of Manchester, England, who was a stranger to the solitudes of primitive forests, and almost a stranger to weird experiences, the silent advance was a mystery. And it began to be a hateful one; for he had not even the poor explanation of it which has been given in this record.

It was only his third night in Maine wilds; and I fear that his friend Cyrus, when inviting him to join in the jacking excursion, had refrained from explaining the canoe mystery, mischievously promising himself considerable fun from the English lad's bewilderment.

Neal's hearing was strained to catch any sound of big game beating about amid the bushes on shore or splashing in the water, but none reached him. The night seemed to grow stiller, stiller, ever stiller, as they glided towards the head of the pond, until the dead quiet started strange, imaginary noises.

There was a pounding as of dull hammers in his ears, a belling in his head, and a drumming at his heart.

He was tortured by a wild desire to yell his loudest, and defy the brooding silence.

Another — a midnight watchman — broke it instead.

"Whoo-ho-ho-whah-whoo!"

It was the thrilling scream of a big-eyed owl as he chased a squirrel to its death, and proceeded to banquet in unwinking solemnity.

"Whoo-ho-ho-whah-whoo!"

Neal started, — who wouldn't? — and joggled the canoe, thereby nearly ending the night hunt at once by the untimely discharge of his rifle.

He had barely regained some measure of steadiness, though he felt as if needles were sticking into him all over, when at last there was a crashing amid the bushes on the right bank, not a hundred yards distant.

Noiselessly as ever the canoe shot around, turning the jack's eye in that direction. A minute later a magnificent buck, swinging his antlers proudly, dashed into the pond, and stooped his small red tongue to drink, licking in the water greedily with a soft, lapping sound.

Neal silently cocked his rifle, almost choking with excitement; then paused for a few seconds to brace up and control the nervous terrors which had possessed him, before his eye singled out the spot in the deer's neck which his bullet must pierce. But he found his operations further delayed; for the animal suddenly lifted its head, scattered feathery spray from its horns and hoofs, and retired a few steps up the bank.

In its former position every part of its body was visibly outlined under the silver light of the jack. Now a successful shot would be difficult, though it might be managed. The boy leaned slightly forward, trying to hold his gun dead straight and take cool aim, when the most curious of all the curious sensations he had felt this night ran through him, seeming to scorch like electricity from his scalp to his feet.

From the stand which the deer had taken,

its body was in shadow. All that the sportsman could discern were two living, glowing eyes, staring — so it appeared to him — straight into his, like starry search-lights, as if they read the death-purpose in the boy's heart, and begged him to desist.

It was all over with Neal Farrar's shot. He lowered his rifle, while the speech, which could no longer be repressed, rattled in his throat before it broke forth.

"I'll go crazy if I don't speak!" he cried.

At the first word the buck went scudding like the wind through the forest, doubtless vowing by the shades of his ancestors that he never would stand to gaze at a light again.

"And — and — I can't shoot the thing while it's looking at me like that!" the boy blurted out.

"You dunderhead! What do you mean?" gasped Cyrus, breaking silence in a gusty whisper of mingled anger and amusement. "You won't get a chance to shoot it or anything else now. You've lost us our meat for to-night."

"Well, I couldn't help it," Neal whispered back. "For pity's sake, what has been moving this canoe? The quiet was enough to set a fellow mad! And then that buck stared straight at me like a human thing. I could see nothing but two burning eyes with white rings round them."

"Stuff!" was the American's answer. "He was gazing at the jack, not at you. He couldn't see an inch of you with that light just over your head. But it would have been a hard shot anyhow, for his nose was towards you, and ten to one you'd have made a clean miss."

"Well," he added, after five minutes of acute listening, "I guess we may give over jacking for to-night. That first cry of yours was enough to set a regiment of deer scampering. I'm only half mad after all at your losing a chance at such a splendid buck. It was something to see him as he stooped to drink in the glare of the jack, a midnight forest picture such as one wants to remember. Long may he flourish! We wouldn't have started out to rid him of his glorious life if we weren't half-starved on flapjacks and ends of pork. Let's get back to camp! I guess you felt a few new sensations to-night, eh, Neal Farrar?"

## CHAPTER II.

### A SPILL.—OUT.

INDEED, shocks and sensations seemed to ride rampant that night in endless succession; a fact which Neal presently realized, as does every daring young fellow who visits the Maine wilderness for the first time, whatever be his object.

Ere turning the canoe towards home, Cyrus drove it a few feet nearer to shore, again warily listening for any further sound of game. Just then another wild, whooping scream cleft the night air; and, on looking towards the bank, Neal beheld his owlship, who had finished the squirrel, seated on an aged windfall,[1] one end of which dipped into the water.

[1] A forest tree which has been blown down.

The gray bird on the gray old trunk formed a second thrilling midnight picture, but at this moment young Farrar was in no mood for studying effects. He felt rather unstrung by his recent emotions; and, though he was by no means an imaginative youth, he actually took it into his head half seriously that the whooping, hooting thing was taunting him with making a failure of the jacking business. Without pausing to consider whether the owl would furnish meat for the camp or not, he let fly at him suddenly with his rifle.

The fate of that ghostly, big-eyed creature will be forever one of those mysteries which Neal Farrar would like to solve. Whether the heavy bullet intended for deer laid him open — which is improbable — or whether it didn't, nobody had a chance to discover. Being unused to birch-bark canoes, the sportsman gave a slight lurch aside after he had discharged his leaden messenger of death, startled doubtless by the loud, unexpected echoes which reverberated through the forest after his shot.

"Hold on!" cried Cyrus, trying to avert a ducking by a counter-motion. "You'll tip us over!"

Too late! The birch skiff spun round,

rocked crazily for a second or two, and keeled over, spilling both its occupants into the black and silver water of the pond.

Of course they ducked under, and of course they rose, gurgling and spluttering.

"You didn't lose the rifle, Neal, did you?" gasped the American directly he could speak.

"Not I! I held on to it like grim death."

"Good for you! To lose a hundred-and-fifty-dollar gun when we're starting into the wilds would be maddening."

Then, just because they were extremely healthy, happy, vigorous fellows, whose lungs had been drinking in pure, exhilarating ozone and fragrant odors of pine-balsam and were thereby expanded, they took a cheerful view of this duck under, and made the midnight forest echo, echo, and re-echo, with peals and gusts and shouts of laughter, while they struggled to right their canoe.

The merry jingles rang on in challenge and answer, repeating from both sides of the pond, until they reached at last the wooded slopes and mighty bowlders of Old Squaw Mountain, a peak whose "star-crowned head" could be imagined rather than discerned against the horizon, near the distant shore from which the hunters had started. Here echo ran riot. It seemed to their excited fancies as if the ghost of Old Squaw herself, the disappointed Indian mother who had, according to tradition, lived so long in loneliness upon this mountain, were joining in their mirth with haggish peals.

The canoe had turned bottom uppermost. On righting it they found that the jack-staff had been dislodged. The jack was floating gayly away over the ripples; its light, being in an air-tight case, was unquenched.

"Swim ashore with the rifle, Neal," said Cyrus. "I'll pick up the jack. Did you ever see anything so absurdly comical as it looks, dodging off on its own hook like a big, wandering eye?"

With his comrade's help young Farrar succeeded in getting the gun across his back, slinging it round him by its leather shoulder-strap; then he struck out for the bank, having scarcely twenty yards to swim before he reached shallow water.

Now, for the first time to-night, the moon shone fully out from her veil of cloud, casting a flood of silver radiance, and showing him a scene in white and black, still and clear as a steel engraving, of a beauty so unimagined and grand that it seemed a little awful. It gave him a sudden respect for the unreclaimed, seldom-trodden region to which his craving for adventure had brought him.

The outline of Old Squaw Mountain could be plainly discerned, a dark, towering shape against the horizon. A few stars glinted like a diamond diadem above its brow. Down its sides and from the base stretched a sable mantle of forest, enwrapping Squaw Pond, of which the moon made a mirror.

"My! I think this would make the fellows in Manchester open their eyes a bit," muttered Neal aloud. "Only one feels as if he ought to see some old Indian brave such as Cyrus tells about, — a Touch-the-Cloud, or Whistling Elk, or Spotted Tail, come gliding towards him out of the woods in his paint and feather toggery. Glad I didn't visit Maine a hundred years ago, though, when there'd have been a chance of such a meeting."

Still muttering, young Farrar kicked off his high rubber boots, and dragged off his coat. He proceeded to shake and wring the water from his upper garments, listening intently, and glancing half expectantly into the pitch-black shadows at the edges of the forest, as if he might hear the stealthy steps and see the savage form of the superseded red man emerge therefrom.

"Ugh! I mind the ducking now more than I did a while ago," he murmured. "The water wasn't cold. Why, we bathed at the other end of the pond late last evening! But these wet clothes are precious uncomfortable. I wish we were nearer to camp. Good Gracious! What's that?"

He stood stock-still and erect, his flesh shrinking a little, while his drenched flannel shirt clung yet more closely and clammily to his skin.

A distant noise was wafted to his ears through the forest behind. It began like the gentle, mellow lowing of a cow at evening, swelled into a quavering, appealing crescendo cadence, and gradually died away. Almost as the last note ceased another commenced at the same low pitch, with only the rest of a heart-beat between the two, and surged forth into a plaintive yet tempestuous call, which sank as before. It was followed by a third, terminating in an impatient roar. The weird solo ran through several scales in its performance, rising, wailing, booming, sinking, ever varying in expression. It marked a new era in Neal's experience of sounds, and left him choking with bewilderment about what sort of forest creature it could be which uttered such a call.

He began to get out some bungling description when Cyrus joined him shortly afterwards, but the American had had a lively time of it while recovering his jack-light and righting the canoe on mid-pond. He was in no mood for explanations.

"Keep the yarn, whatever it is, till tomorrow, Neal," he said. "I didn't hear anything special. Perhaps I was too far away. I'm so wet and jaded that I feel as limp as a washed-out rag. Let's get back to camp as fast as we can."

## CHAPTER III.

#### LIFE IN A BARK HUT.

IT was two o'clock in the morning when the tired, draggled pair stumbled ashore at the place where they embarked, hauled up their birch skiff, leaving it to repose, bottom uppermost, under a screen of bushes, and then stood for some minutes in deliberation.

"I'm sure I hope we can find the trail all right," said Cyrus. "Yes, I see the blazes on the trees. Here's luck!"

He had been turning the jack-lamp on either side of him, trying to discover the "blazes," or notches cut in some of the trunks, which marked the "blazed trail"—in other words, the spotted line through the otherwise trackless forest, which would lead him whither he wanted to go.

It required considerable experience and unending watchfulness to follow these "blazes"; but young Garst seemed to have the instinct of a true woodsman, and went ahead unfalteringly, if vigilantly, while Neal followed closely in his tracks.

After rather a lengthy trudge, they reached a point where the ground sloped gently upward into a low bluff. Still keeping to the trail, they ascended this eminence, finding the forest not so dense, and the walking easier than it had been hitherto. Gaining the top, they emerged upon an open patch, which had been cleared of its erect, massive pines, and the long-hidden earth laid bare to the sky by the lumberman's axe.

Here the eagerly desired sight — that sight of all others to the tired camper; namely, the camp itself, with its cheery, blazing camp-fire — burst upon their view, sheltered by a group of sapling pines, which had grown up since their giant brothers went to make timber.

Now, a Maine camp, as every one knows, may consist of any temporary shelter you choose to name, according to the tastes and opportunities of its occupants, from a fair white canvas home to a log cabin or a hastily erected canopy of spruce boughs. In the present instance it was a "wangen," or hut of strong bark, such as is sometimes used by lumbermen to rest and sleep in when they are driving their floats of timber down one of the rivers of this region to a distant town, which is a centre of the lumber trade.

Cyrus and Neal were making across the clearing in the direction of the camp-fire with revived spirits, when the American suddenly grabbed his friend by the arm, and drew him behind a clump of low bushes.

"Hold on a minute!" he whispered. "By all that's glorious, there's Uncle Eb singing his favorite song! It's worth hearing. You never listened to such music in England."

"I don't suppose I ever did," answered Neal, suppressed laughter making him shake.

Upon a gray pine stump, beside the blaze, which he was feeding with a hemlock bough, sat a battered-looking yet lively personage. Had he been standing upright upon the remnant of trunk, he would certainly, in the bright but changeful firelight, have deceived an onlooker into believing him to be a continuation of it; for the baggy tweed trousers which he wore on his immense legs, and which partially hid his loose-fitting brogans, or woodsman's boots, his thick, knitted jersey, his mop of woolly hair, with the cap of coon's fur that adorned it, were a striking mixture of grays, all bordering upon the color of the stump. His skin, however, was a fine contrast, shining as he bent towards the flame like the outside of a copper kettle. In daylight it would be three shades darker, because the thick coral lips, gleaming teeth, and prominent, friendly eyes of the individual, betrayed him to be in his own words, "a colored gen'leman;" that is, a full-blooded negro, and a free American citizen.

Beside him, squatting upon his haunches and wagging his shaggy tail, was a good-sized dog, not of pure breed, but undoubtedly possessed of fire and fidelity, as was shown by the eye he raised to his master. His red coat and general formation showed that his father had been an Irish setter, though he seemed to have other and fiercer blood in his veins, mingling with that of this gentle parent.

To him the negro was chanting a war-song, — some lines by a popular writer which he had found in an old newspaper, and had set to a curious tune of his own composition, rendering the performance more inspiriting by sundry wild whoops, and an occasional whacking of his teeth together.

Here are two verses, under the influence of which the dog worked himself up to such excitement that he seemed to feel the ghosts of rabbits slain — for he could smell no live ones — hovering near him : —

> "I raise my gun whar de rabbit run —
>   Ketch him, Tiger, ketch him!
> En de rabbit say :
>   'Gimme time ter pray,
> Fer I ain't got long fer to stay, to stay!'
>   Oh, ketch him, Tiger, ketch him!
>
> "Ketch him, oh, ketch him!
>   Run ter de place en fetch him!
> De bell done chime
> Fer de breakfast time —
>   Oh, ketch him, Tiger, ketch him!"

"If there are any more verses, Uncle Eb, keep them until we've had supper, or breakfast, or whatever you like to call a meal at this unearthly hour. I'm so hungry that I could chew nails!" cried Cyrus, springing from behind the bushes, and reaching the camp-fire with a few strides, Neal following him.

"Sakes alive! yonkers; is dat you?" cried the darkey, uprearing his gray figure. "I'se mighty glad to see you back. Whar's yer meat? Left it in de canoe mebbe? De buck too big to drag 'long to camp — eh?"

There was a wicked rolling of Uncle Eb's eyes while he spoke. Evidently from the looks of the sportsmen he guessed immediately what had been the result of their excursion.

"No luck and no buck to-night!" answered Garst. "But don't roast us, Uncle Eb. Get us something to eat quicker than lightning or we'll go for you — at least we would if we weren't entirely played out. It isn't everybody who can manage a hard shot as cleverly as you do, when he can only see the eyes of an animal. And that was the one chance we got."

No man living ever heard a further word from Cyrus as to how his English friend bore the scares of a first night's jacking.

"Ya-as, dat's a ticklish shot. Most folks is skeered o' trying it," drawled out Ebenezer Grout, a professional guide as well as "colored gen'leman," familiarly called by visitors to this region who hired the use of his hut and his services, "Uncle Eb."

"There's some comfort for you," whispered Cyrus slyly into Neal's ear. Aloud he said, addressing the guide, "We had a spill-out, too, as a crown-all. I'm mighty glad that this is the second of October, not November, and that the weather is as warm as summer; otherwise we'd be in a pretty bad way from chill. I feel shivery. Hurry up, and get us some steaming hot coffee and flapjacks, Uncle Eb, while we fling off these wet clothes. The trouble is we haven't got any dry ones."

"Hain't got no oder suits?" queried the woodsman. "Den go 'long, boys, and rig yerselves up in yer blankets. Ye can pertend to be Injuns fer to-night. Like enough dis ain't de worst shift ye'll have to make 'fore ye get out o' dese parts."

As the draggled pair were making towards the hut, which stood about six feet from the fire, to follow his advice, its bark door was suddenly pushed wide open. Forth stepped, or rather staggered, another boy, younger and shorter than Neal. His tumbled fair hair was here and there adorned with a green pine-needle, which was not remarkable, considering that he had just arisen from a bed of pine boughs. Sundry others were clinging to the surface of the warm, fleecy blankets in which he was wrapped, and his feet were thrust into a pair of moccasins. He had the appearance and voice of a person awaking from sound sleep.

"I say, you fellows, it's about time you got back!" he said, rubbing his heavy eyes, and addressing the hunters. "I hope you've had some luck. I dreamt that I was smacking my lips over a venison steak."

"Smack 'em w'en you git it, honey!" remarked Uncle Eb, while he mixed a plain batter of flour, baking-powder, and cold water, which he dropped in big spoonfuls on a frying-pan, previously greased, proceeding to fry the mixture over his camp-fire.

The thin, round cakes which presently appeared were the "flapjacks" despised by Cyrus as insufficient diet.

Without waiting to answer the new boy's greeting, the hunters had disappeared into the bark shanty. When next they issued forth they were rigged up Indian fashion in moccasins and blankets, the latter being doubled and draped over their underclothing, — of which luckily they had a dry supply, — and gathered round their waists with leather straps. Knitted caps, usually worn when sleeping, adorned their heads.

"You see, we followed Dol's example and your advice, Uncle Eb," said Cyrus, as they seated themselves by the camp-fire. "And I tell you these make tip-top dressing-gowns when you're feeling a little bit chilly after a drenching. We didn't bring along a second suit of tweeds for the simple reason that we mean to do some pretty rough tramping with our packs on our backs, and then a fellow is likely to grumble at any unnecessary pound of weight he carries."

"Shuah — shuah!" assented Uncle Eb.

"And that is why we left our fishing-rods behind," continued Garst. "You see, our main object this trip is neither hunting nor fishing. But a creel of gamey trout from Squaw Pond would come in handy now to replenish our larder."

"Wal, I b'lieve I'll fix up a rod to-mo-oh an' hook a few, fer de pork's givin' out. Hain't got mich use fer trout meself. Dey's kind o' tasteless eatin' if a man can git a bit o' fat coon or a fatty [hare], let 'lone ven'zon. Pork's a sight better'n 'em to my mind."

While Uncle Eb was giving his views on food, he was hurriedly "bilin'" coffee, frying unlimited flapjacks, and breaking up some crystal cakes of maple sugar, which he melted into a sirup, and poured over them.

> "De bell done chime
> Fer de breakfast time!"

he shouted gleefully when all was accomplished. "Heah, yonkers! I guess we may call dis meal breakfast jest as well as not, fer it's neah to dawn now."

And the trio fell to voraciously, as he handed them each a steaming tin mug and an equally steaming plate. The newly awakened youngster, who had been cuddling his head sleepily against Neal's shoulder (a glance showed that they were brothers), had clamored for his share of the banquet.

"You haven't been lonely, Dol, I hope, have you?" said Cyrus, as a whole flapjack, doubled over and drenched in sirup, disappeared down his capacious throat.

"Not I," answered Dol (Adolphus Farrar, ladies and gentlemen), shutting and opening a pair of steel-gray eyes with a sort of quick snap. "Uncle Eb and I sat by the fire until twelve o'clock. He sang songs, and told tip-top stories about coon hunts. I tell you it was fun! I'd rather see a coon hunt than go out at night jacking, especially if I got a ducking instead of a deer, like some bungling fellows I know."

"Don't be saucy, Young England, or I'll go for you when I've finished eating," laughed Cyrus good-humoredly. "Who told you what we got?"

Dol winked at Uncle Eb, who had, indeed, entertained him with giggling jokes about the unsuccessful hunters while they were stripping off their wet garments.

Adolphus, being the youngest of the camping-party, was favored with the softest pine-bough bed and the best of the limited luxuries which the camp possessed, with unlimited nicknames, — from "Young England" to "Shaver" or "Chick," according to the whims of his comrades.

"Say, Uncle Eb, we're having a fine old time to-night — all sorts of experiences! I guess you may as well finish that song we interrupted while we're finishing our meal."

"All rightee, gen'lemen!" answered the jolly guide and cook.

The dog Tiger had retreated to the back of the camp-fire, where he lay blissfully snoozing; but at a booming "Whoop-ee!" from his master, which formed a prelude to the following verses, he shot up like a rocket, and manifested all his former signs of excitement.

> "Dey's a big fat goose whar de turkey roos' —
>   Ketch him, Tiger, ketch him!
> En de goose — he say,
>   'Hit'll soon be day,
> En I got no feders fer ter give away!'
>   Oh, ketch him, Tiger, ketch him!
>
> "Ketch him, oh, ketch him,
> Run ter de roos' en fetch him!
> He ain't gwine tell
> On de dinner bell —
>   Ketch him, Tiger, ketch him!"

"Scoot 'long to bed now, you yonkers, or ye'll look like spooks to-mo-oh! Hit's day a'ready," cried the singer directly he had whooped out his last note.

And the "yonkers," nothing loath, for they had finished their repast, sprang up to obey him.

"Isn't it a comfort that we haven't any trouble of undressing and getting into our bedclothes, fellows?" Cyrus said, as they reached the wangen, and prepared to throw themselves upon the fragrant camp-bed of fresh green pine-boughs, which made the bark hut smell more healthily than a palace.

The natural mattress was wide enough to accommodate three. The boughs were laid down in rows with the under side up, and overlapped each other. To be sure, an occasional twig might poke a sleeper's ribs, but what mattered that? To the English boys especially — having the charm of entire novelty — it was a matchless bed, wholesome, restful, and rich with balsamic odors hitherto unknown.

The trio were stupidly tired; but on the American continent no happier or healthier youths could have been found.

It had, indeed, been a night big with experiences; and there was one still to come, which, to Neal Farrar at any rate, was as novel as the rest. He had thrown himself upon his bough couch, too weary to offer anything but the gladness of his heart for worship, when Cyrus touched his arm.

"Look there!" he said. "If a fellow could see that without feeling some sensations go through him which he never felt before, he wouldn't be worth much!"

He pointed through the open door of the hut at the sky above the clearing, over which was stealing a pearly hue of dawn, shot with a tinge of rosy light, like the fire in the heart of an opal.

This made a royal canopy over the towering head of Old Squaw Mountain,—near by now and plainly visible,—which had not yet lost its starry diadem, though the gems were paling one by one. The shoulders of the peak wore a mantle of purple, and the forest which clothed its bulk was changing from the blackness of a mourning robe to the emerald green of a sea-nymph's drapery.

The shutters of Night were rolling back, and young Day was stepping out to cast her first smile on a waiting earth.

As the watchers in the hut caught that smile, every thought which rose in them was a daybreak song to the God who is light, and the secret of every dawning.

With the day-smile kissing their faces they fell asleep, feeling that they were wrapped in the embrace of the invisible King.

## CHAPTER IV.

#### WHITHER BOUND?

"WHERE from? Whither bound?" It is not often that a man or boy burns to put these questions — which ships signal to each other when they pass upon the ocean — to some individual who hurries by him on a crowded thoroughfare, whose name perhaps he knows, but whose hand he has never clasped, of whose thoughts, feelings, and capabilities he is ignorant.

But just let him meet that same fellow during a holiday trip to some wild sea-beach or lonely mountain, let an acquaintance spring up, let him observe the habits of the other traveller, discovering a few of his weak points and some of his good ones, and then he wishes to ask, "Where do you hail from? Whither are you bound?"

Therefore, having encountered three fairly good-looking, jovial, well-disposed young fellows amid the solitudes of a Maine forest, having spent some eventful hours in their company, learning how they behaved in certain emergencies, it is but natural that the reader should wish to know their ordinary occupations, with their reasons for venturing into these wilds, and the goal they wish to reach, before he journeys with them farther.

Just at present, being fast asleep, dreaming, and — if I must say it — snoring like troopers, upon their mattresses of pine boughs, they are unable to give any information about themselves. But the friend who has been authorized to record their travels will be happy to satisfy all reasonable curiosity.

To begin, then, with the "boss" of the party, Cyrus Garst, the writer would say that he is a student of Harvard University, and a brainy, energetic, robust son of America. Among his college classmates he is regarded as a bit of a hero; for, in spite of his comparative youth, he is an enterprising traveller and a veteran camper, whose camp-fire has blazed in some of the wildest solitudes of his native land. For his hobby is natural history, and his playground the " forest primeval," where he studies American animals amid the lonely passes which they choose for their lairs and beats.

Every year when Harvard's learned halls are closed for the long summer vacation, — sometimes at other seasons too, — he starts off on a trip to a wilderness region, with his knapsack on his back, his rifle on his shoulder, and often carrying his camera as well.

Once in a while he has been accompanied by a bosom friend or two. More frequently he has gone alone, hiring the services of a professional guide accustomed to the locality he visits. Now, such a guide is the indispensable figure in every woodland trip. He is expected to supply the main part of his employer's camp " kit " ; namely, a tent or some shelter to sleep under, cooking utensils, axes, etc., as well as a boat or canoe if such be required. And this son of the forest, whose foot can make a bee-line to its destination through the densest wooded maze, is not only leader, but cook and general-utility man in camp as well. The guide must be equally grand-master of paddle, rifle, and frying-pan.

For these tireless woodland heroes Cyrus

Garst has a general admiration. He has always agreed with them famously — save on one point; and he has never had to shorten his wanderings for fear of lengthening their fees. For Cyrus has a millionnaire father in the Back Bay of Boston, who is disposed to indulge his whims.

The one point of variance is this: while all guides admire young Garst as a crack shot with a rifle, he frequently dumfounds them by letting slip stunning chances at game, big and little. They call him "a queer specimen sportsman," — understanding little his love for the wild offspring of the woods, — because he never uses his gun save when the bareness of his larder or the peril of his own life or his chum's demands it.

Nevertheless, feeling the need of fresh meat, the naturalist was for the moment hotly exasperated because his English comrade, Neal Farrar, missed even a poor chance at a buck during the midnight excursion on Squaw Pond.

His friends are proud of stating that up to the present Cyrus had proceeded well in his friendly acquaintance with wild creatures, his desire being to study their habits when alive rather than to pore over their anatomy when dead. And he has always reaped a plentiful harvest of fun during his trips, declaring that he has "the pull over fellows who go into the woods for killing," seeing that he can thoroughly enjoy the escape of a game animal if he can only catch a sight of it, and perceive how its pluck or cunning enables it to baffle pursuing man. There are those who call Cyrus a sportsman of the best type. Perhaps they are right.

Yet in the year of our story, when he had just attained his majority, this student of forest life is still unsatisfied, because he has not been able to obtain a good view of the behemoth of American woods, the *ignis fatuus* of hunters, — the mighty moose.

Once only, when paddling on a still pond with his experienced guide for company, the latter suddenly closed the slide of the jack-lamp, hiding its light. At the same moment a dark, splendid monster, tall as a horse and swinging a pair of antlers five feet broad, suddenly appeared upon the bank, near to which the canoe lay in black shadow. The hunters dared not breathe. It was at a season of year when the Maine law exacts a heavy fine for the killing of a moose; and even the guide had no desire to send his bullets through the law, though he might have riddled the game without compunction.

For a minute or two the creature halted at the pond's brink, magnified in the mirror of moonlit water into a gigantic, wavering shape. Then with slow, solemn tread he walked along the bank ahead, gave a loud snort something like the snort of a war-horse, made a crunching, chopping noise with his jaws, resembling the sound of a dull axe striking against wood, plunged into the lake, and swam across to the opposite shore.

"If we had fired, he might have come for us full tilt," whispered the guide so softly that his words were like a gliding breath. "And then I tell you we'd have had a narrow squeak. He'd have kicked the canoe into splinters and us out o' time in short order."

"But a moose won't charge unless he's attacked, will he?" asked Cyrus, later in the night, when a couple of quacking black ducks which had received a dose of lead were lying silent at his feet, and the hunters were returning to camp with food.

"Not often," was the reply. "Only at this time o' year, if they've got a mate to defend, you can't say for sure what they'll do. They won't always fight either, even if they're wounded, when they can get a chance to bolt. But a moose, if he has to die, will be sure to die game, with his face to his enemy; and so will every wild animal that I know. I've even seen a shot partridge flutter up its feathers like a game-cock at the fellow who dropped it."

Well, this memorable glimpse of his mooseship was obtained in the year before our story. And now, in the beginning of October, young Garst was off into Maine wilds again, having arranged to "do" the forest thoroughly after his usual fashion, seeing all he could of its countless phases of life, and finally to meet this same guide — a dare-devil fellow who was reported to have had adventures in moose-hunting such as other woodsmen did not dream of — at a log camp far in the wilderness. Thence they could proceed to solitudes where the voice of man seldom echoed, where the foot of man rarely trod, and where moose signs were pretty sure to be found.

But there was one very unusual feature in his present expedition. The student of nature, who generally started forth alone, was this year, owing to a freak of fate and to his natural good-nature, accompanied by two English lads.

Early in the summer of this same year, Francis Farrar, a wealthy cotton-merchant of Manchester, England, visited America on a business-trip, and became the guest of Cyrus's father. He brought with him his two sons, Neal, aged sixteen and a half, and Adolphus, familiarly called Dol, who was more than a year younger.

Both boys had been at a large public school, and physically, as well as mentally, were well developed. They were accustomed to spending long vacations with their father at wild spots on the seashore, or amid mountains in England and Scotland. They could tirelessly do a sixty-mile spin on their "wheels," were good football players, excellent rowers, formed part of the crew of their father's yacht, could skilfully handle gun and fishing-rod, but they had never camped out.

They knew none of the delights of sleeping in woodland quarters, with only a canvas or bark roof, or perhaps a few spruce boughs, between them and the sky —

> "While a music wild and solemn
> From the pine-tree's height
> Rolls its vast and sea-like volume
> On the wind of night."

Small wonder, then, that when they heard Cyrus Garst tell of his camping excursions, of his jolly times, long tramps, and hair-breadth escapes, their hearts swelled with a tremendous longing to accompany him on the trip into northern Maine which he was then projecting for the following October.

Now, Cyrus at the first start-off conceived a liking for these English fellows, to whom, for his father's sake, he played the part of genial host. With a lordly recognition of his superior years he pronounced them "first-rate youngsters, with lots of snap in them." And as the acquaintance progressed, Neal Farrar, with his erect figure, broad chest, musical voice, and wide-apart gray eyes, — so clear and honest that their glance was a beam, — proved a personage so likable that the student adopted him as "chum," forgetting those five years which had been a gulf between them.

Dol, whose eyes were of a more steely hue than his brother's, striking fire readily and showing all manner of flinty lights, who had a downright talent for mimicry, and a small share of juvenile self-importance, came in for regard of a more indulgent and less equal nature.

Directly he got an inkling of the desire for a forest trip which stirred in the boys' breasts, making them yearn all day and toss all night, Cyrus gave them both a cordial invitation to accompany him into Maine. Mr. Farrar did not purpose returning to Europe till midwinter. His consent was easily obtained. He presented each of his sons with a new Winchester repeating rifle, with which they practised diligently at a target ere the eventful day of the start dawned, though their leader emphatically insisted that the prime pleasures of the trip were not to be looked for in the slaughter done by their hands.

Wearing the camper's favorite dress of stout gray tweed, the trio left Boston on a lovely September evening towards the close of the month, taking a fast night train for Maine, brimful of enthusiasm about the wild woods and free camp-life. The hue of their clothes was chosen with a view to making their figures resemble the forest trunks, so that they would be less likely to attract the notice of animals, and might get a chance to creep upon them undetected.

About their waists were their ammunition belts, with pouches well stocked. Their large knapsacks contained blankets, moccasins, and various other necessaries of a camper's outfit, including heavy knitted jerseys for chill days and nights, and rubber boots reaching high on the legs for wear in wading and traversing swampy tracts.

About twenty-four hours later they dropped off the rattling, jingling stage-coach which bore them over the latter part of their journey, at the flourishing village of Greenville, on the borders of the Maine wilds.

Here they were greeted by a view, the loveliness of which made the English boys, who had never looked on it before, experience strange heart-leaps.

A magnificent sheet of water nearly forty miles long and fourteen broad lay before them, studded with islands, girt with evergreen forests and wooded peaks. Under the rays of the setting sun its bosom was shot with arrows of pale, quivering gold. Banners of gold and flame-color floated over the crests of the hills, flinging streamers of light down their emerald sides.

"Fellows, there is Moosehead Lake; and I guess you'll find few lakes in America or elsewhere that can beat it for beauty," said Cyrus, with a patriotic thrill in his voice, for he had a feeling that he was doing the honors of his country.

His English comrades were warm with admiration, and here, in view of the forest-land which was their El Dorado, tingled with anticipation of the unknown.

The three rested that night at Greenville, and began their tramping on the following morning. They trudged a distance of seven miles or so to the camp of Ebenezer Grout, which, as Garst knew, was situated between Squaw Pond and Old Squaw Mountain, the latter being one of the finest peaks near Moosehead Lake.

"Uncle Eb" was an old acquaintance of Cyrus's, a dusky, lively woodsman, who spent a great part of the year in his lone bark hut, with his dog Tiger for company. He subsisted chiefly on what he brought down with his rifle, and sometimes earned three dollars a day for guiding tourists up Old Squaw or through the adjacent forests.

He was not an ambitious hunter, and rarely pushed far into the solitudes of the wilderness in search of moose or other big game. A coon hunt was to him the climax of all fun. It was chiefly with a hope that his comrades might enjoy some novel entertainment of this

kind that Cyrus made his first stoppage at Uncle Eb's camp, purposing to sojourn there for a few days.

He was not disappointed.

The stupidly tired trio had slept for about two hours, while the reader has been receiving information second-hand about their past and future, when a scratching, scraping, boring noise on the outside of their bark roof temporarily disturbed their slumbers. Dol called out noisily, and, as was the way of that youngster on sundry occasions, talked some gibberish in his sleep. The scraping instantly ceased.

A renewed and blissful season of snoring. Another awakening. More music on the roof, evidently caused by the claws of some wild animal, while each of the campers was startled by a loud " Cluck!"

"Lie still, fellows! Don't budge. Let's see what the thing is," breathed Cyrus in a peculiarly still whisper which he had learned from his moose-hunting guide of whom mention has been made.

Dead silence in the hut. Redoubled scraping and rattling above, with a scattering of bark chips.

Then light appeared through a jagged hole

just over a string which was stretched across one corner of the cabin, and from which dangled sundry articles of camp bric-a-brac, mostly of a tinny nature, with Uncle Eb's last morsel of pork.

"By all that's glorious! it's a coon," breathed Cyrus, but so softly that his companions did not hear.

As for the two Farrars, they were working up to such a heat of excitement that they felt as if life were now only beginning. They had heard of the thievish raids made by the black bear on unprotected camps, and of his special fondness for pork. Not knowing that there was no chance of an encounter with Bruin so near to civilization as this, they peered at that hole in the roof, expecting every moment to see a huge, black, snarling snout thrust through it.

It was a pointed gray muzzle which warily appeared instead — appeared and disappeared on the instant. For at this crisis Tiger's shrill bugle-call resounded without, giving warning of an attack on the camp. The thing, whatever it was, scrambled from the roof, and with a strange, shrill cry of one note made towards the woods. The dog followed it, barking for all he was worth.

Now, too, Uncle Eb's booming "Whoopee!" was heard.

The hardy old woodsman, after his visitors had gone to roost, instead of stretching himself as usual upon his pine mattress, had started off, accompanied by Tiger, to visit some traps which he had set in the forest, hoping to catch a marten or two. He took the precaution of closing the door of the hut when he saw that its inmates were soundly sleeping, thinking meanwhile, that, as day was dawning, there was little chance of any wild "critter" coming round the camp during his absence.

But a greedy raccoon, which had been prowling near in the woods during the night, and had been tantalized to desperation by the smell of the late meal, especially by the odor of flapjacks frying in pork fat, had stolen from cover after the departure of his natural enemy, the dog.

Finding the coast clear and the camp unguarded, he made himself quietly at home, rooted among some potato parings which the guide had thrown aside a day or two before, devoured a cold flapjack, and cleaned the camp frying-pan as it had never been cleaned before, with his tongue. But his appetite was whetted, not glutted. Scent or instinct told him that pork, molasses, and other eatables were hidden in the bark hut. Here was a golden opportunity for Mr. Coon. No one molested him. Meditating a feast, he climbed to the roof, and began cautiously to scrape off portions of the bark. The rising sun ought to have warned him back to forest depths; but he persisted in his scratching, repeating now and again a satisfied cluck.

His hole was made. His keen nose told him that pork was almost within reach, when the bugle-call of his enemy — Tiger's challenging bark — smote upon his ear. Guide and dog were opportunely returning to camp.

Of course, as soon as the marauder scrambled off the roof, Cyrus and the boys sprang from their couch. Barefooted, and in night costume, they were already at the door of the hut before Uncle Eb was heard booming, —

"Boys! Boys! Tumble out — tumble out! Dere's a reg'lar razzle-dazzle fight goin' on heah. Tiger's nabbed de coon."

## CHAPTER V.

#### A COON HUNT.

A RAZZLE–DAZZLE fight it surely was! On one side of the camp, between the camping-ground, which Uncle Eb had cleared with many a backache, and the woods, was a narrow strip covered with a stunted, prickly growth of wild raspberry bushes and tiny cherry-trees. These had sprung up after the pines had been cut down, as soon as the sun peeped at the long-hidden earth.

Into it the bare-legged trio dared not venture, knowing that they would get a worse scratching and tearing than if the coon itself mauled them.

But they could see and hear a whirling, howling, clawing, spitting, rough-and-tumble conflict going on in the midst of this miniature jungle.

"Whew! Whew!" gasped Cyrus. "Here's your first sight of a wild coon, boys. I wish to goodness it had been a different sight, but I suppose he must pay for his thieving."

"Tiger'll make him do dat. Bet yer life he will! He's death on coons, if ever a dog was," yelled Uncle Eb, gambolling with excitement, his eyes bulging and widening until they looked like oysters on the shell.

The soft, battered, gray felt hat which replaced his fur cap in the daytime surged off his gray wool, and frisked gently away towards the camp-fire. There, coming in contact with a red ember, it scorched and shrivelled into smoking, smelling ashes, all unnoticed in the tumult of the fight.

Whirling round and round, now under, now over, dog and coon rolled presently forth from the bushes, nearer to the feet of the spectators. Then Neal and Dol could get a clearer view of the strange animal. A breeze of exclamations came from them, mingling with the yelping, snarling, and clucking of the combatants.

"Good gracious! Look at the stout body and funny little legs of the fellow!"

"Doesn't he fight like a spitfire?"

"I'm glad he's not clawing me!"

"He's not much like any picture of a raccoon I ever saw in a Natural History!"

"I guess he wouldn't resemble them greatly, especially in that attitude, Dol," said Cyrus, as soon as there was a lull in the boys' comments.

The raccoon had now rolled on his back, and was fighting so fiercely with teeth and claws that a despairing cry broke from Uncle Eb, —

"Yah! He's makin' Tiger's wool fly!"

It was then that the old guide began to deliberate about rushing forward and despatching his coonship with the butt end of his rifle. Cyrus would gladly have stopped the tussle long before, for there was too much savagery about it to suit him; but he could only have done so by stunning or killing one of the combatants.

A heart-rending howl from Tiger. The coon had caught him by his lower jaw. Uncle Eb, clutching his empty rifle like a club, was starting to the rescue, when the dog with a sudden, desperate jerk freed himself. Mad with rage and pain, he tried to seize the raccoon's throat. But his enemy managed to elude the strangling grip, and getting on his feet, again caught Tiger, this time by the cheek, causing another agonizing yelp.

Now, however, the undaunted dog whirled round and round with such rapidity as to make Mr. Coon relax his hold, and, gathering all his strength, flung the wild animal off to a distance of several feet.

Probably the raccoon felt that he had enough of the conflict, and was doubtful about its final issue. He seized the chance for escape. While the spectators gasped with excitement, they beheld him, with his head doubled under his stomach, roll over and over like a huge gray India-rubber ball, until he reached the nearest tree, which happened to be one of the young pines that shaded the camp. Quick as lightning he climbed up its trunk, uttering a second shrill, far-reaching cry of one note.

"Listen! Listen, fellows!" cried Cyrus. "That raccoon is a ventriloquist. The cry seemed to come from somewhere far above him. I had a tame coon long ago, and I often heard him call like that. I tell you he's a ventriloquist, and a mighty clever one too.

"The one piercing note was to warn his mate," went on the naturalist, after a moment's pause; "or in all probability, though we have been speaking of the animal as 'he,' it is really a female, for I have heard that peculiar call given more frequently by a mother to warn her cubs."

All that could now be seen of the animal — on whose gender new light had been cast — was a gray ball curled up on a tasselled bough near the top of the pine-tree, and a glimpse of a black nose over the edge of the limb.

"Wal! 'tain't no matter wedder de critter is a male or a fimmale; I'm a-goin' to bring it down from dar mighty quick," said Uncle Eb, fumbling with the cartridge-box which was attached to his broad leather belt, and preparing to load his rifle, while he cast murderous looks aloft.

"No, you don't, then!" said Cyrus hotly. "The creature has fought pluckily, and it deserves to get a fair chance for its life. I'll see that it does too. You oughtn't to be hard on it for liking pork, Uncle Eb."

"Coons will be gittin' into eatin' order soon," murmured the guide, smacking his lips, and handling his gun undecidedly. "Roast coon's a heap better'n roast lamb."

"Well, they're not in eating order yet, and

won't be till next month," answered Garst. "Come, you've got to let this one go, Uncle Eb, to please me."

"Tell ye wot: I'll call Tiger off" (Tiger was alternately licking his wounds and baying furiously for vengeance about the tree which sheltered his enemy), "den, wen de coon finds de place clear, bime-by he'll light down from dat limb, I'll start off de dog, and let 'em finish de game atween 'em."

Cyrus considered for a minute, then decided that on the coon's behalf he might safely accept the compromise.

"Let's get into our clothes, fellows!" he cried to Neal and Dol. "Now we're going to have some fair fun! I guess there won't be any more fighting; and I want you to see how cunningly the raccoon will cheat the dog and escape, if he gets an even chance."

In five minutes the trio were out of their blankets and in their ordinary day apparel. The old guide had hung the wet tweeds to dry by the blazing camp-fire before he started out to visit his traps, carefully stretching them to prevent their "swunking" (shrinking). Thus they were again fit for wear.

A half-hour of waiting ensued, during which every one was on the tiptoe of expectation. They had all withdrawn to some distance from the tree. Uncle Eb had been obliged to drag Tiger away, and was bathing his cuts out of the camp water-bucket in a shady corner. The dog, recognizing that he was a patient, submitted without a growl or budge, until his master, who had been keeping a keen eye on that pine-tree, suddenly loosed him, and started him off afresh with a loud "Whoop-ee!" and a —

"Ketch him, Tiger! ketch him!"

The coon had "lighted down."

Away went the wild creature into the woods. Away after him, went dog, guide, student, and boys, plunging, tumbling, rushing along helter-skelter, with a yell on every lip.

"There he is! See him? That gray ball rolling over and over!" shouted Cyrus. "I'll tell you what, now; he's going to resort to his clever dodge of 'barking a tree.' There never was a general yet who could beat a coon for strategy in making a retreat."

The forest surrounding the eminence on which Uncle Eb's camp was situated consisted mostly of pines, with here and there the brilliant autumn foliage of a maple or birch showing amid the evergreens. The trees down the sides of the hill were not densely crowded, but grew in irregular clumps instead of an unbroken mass. This, of course, afforded a better opportunity for the pursuers to catch glimpses of the fugitive animal.

On finding that it was again chased, the raccoon at first took shelter in a dense thicket of scrub oak, which formed in places a tangled undergrowth. Tiger quickly followed up its trail, and it was driven thence.

Then Cyrus and the boys caught sight of it spinning over and over like a ball, towards a maple-tree with widely projecting limbs and thick foliage; for it knew well that in speed it was no match for the dog, and therefore resorted to a neat little stratagem. The next minute, being hotly pressed, it scrambled up the friendly trunk.

"He's treed again, yonkers! Come on!" shouted the guide, indifferent to the creature's probable gender.

Tiger sat on his haunches at the foot of the maple, setting up a slow, steady bark.

"Keep where you are, fellows! Watch the other side of the tree!" whispered Cyrus, his face twitching with excitement.

In his character of naturalist he had managed to find out more about the coon's various dodges than even the old guide had done.

In breathless wonder the Farrars presently beheld that ingenious raccoon steal along to the end of the most projecting limb on a different side of the tree from the one it had climbed, so that a screen of boughs and the trunk were between it and its adversary.

Then it noiselessly dropped from the tip of the branch to the ground, alighting, like a skilled acrobat, on its shoulders, doubled its pointed black nose under its stomach, and again rolled over and over for a considerable distance, when it got on its short legs and scurried away, while Tiger still bayed at the foot of the maple-tree, thinking the vanished prey was above.

"That's what I called the coon's dodge of 'barking a tree,'" said Cyrus. "Don't you see, when hard pressed, he runs up the trunk, leaving his scent on the bark; then he creeps to the other side under cover of the foliage, and drops quietly to the ground. So he breaks the scent and cheats the dog."

"Good gracious!" exclaimed Neal with an expressive whistle.

"Perhaps it's because of his long gray hairs that he has so much wisdom," Dol suggested.

"A bright idea, Chick!" chuckled the student, tapping the boy's shoulder.

"We keep on speaking of him as 'he' when you said the thing was probably a female," put in Neal.

"That doesn't matter. I'm not certain. Look at old Tiger! He's having fits now that he has discovered how he's been tricked."

The dog was circling out from the tree, with wild, uncertain movements, nosing everywhere. Presently he struck the scent again, and darted off like a streak.

But the raccoon had by this time reached a dark stream of water which coursed through the over-arching forest at the foot of the hill, as if it was flowing through a tunnel. Here this astute animal crossed and recrossed under the gloom of interlocking trees, mid dense undergrowth, until its trail was altogether lost.

Tiger, having further "fits," nosing about, darting hither and thither, venting short, baffled barks, finally gave up in despair.

The pursuing party turned back to camp.

"Did ye ever see ennyting to ekal de cunnin' o' de critter," said Uncle Eb gloomily; "runnin' up dat tree on'y to jump off, so as he'd break de scent an' fool de dog? Ye'll learn a heap o' queer tings in dese woods, chillun, 'fore ye get t'rough," he added, addressing the English lads.

"We've learned queerer things than we ever imagined or dreamed of, already, Uncle Eb," Neal answered.

Meanwhile, Cyrus and Dol had begun to discuss the size of the escaped coon.

"I should think it measured about two feet from the tip of its nose to the beginning of the tail, and that would add ten or eleven inches. Probably it weighed over thirty pounds," said the experienced Garst.

"A fine tail it had too!" answered Dol;" all ringed with black and buff — not black and white as the books say. There was hardly an inch of white about the animal anywhere. Its thick gray hair was marked here and there with black; wasn't it, Cy?"

"Rather with a darker shade of gray, bordering on black. I think old Tiger can testify that the creature had capable teeth; and it possesses a goodly number of them — forty in all; that's only two less than a bear, an animal that might make six of it in size."

"Whew! No wonder it's a good fighter!" ejaculated Dol.

"But the funniest of the coon's or — to give the animal its proper name — the raccoon's funny habits is, that while it eats anything and everything, it souses all meat in water before beginning a feed. That's what it would have done with our bit of pork,— dragged it to a stream, and washed it well before swallowing a morsel.

"I caught glimpses of a raccoon chasing a jack-rabbit in this very section of the woods, last year," went on the student, seeing that Dol was breathlessly listening. "The big animal killed the little one under a dead limb; and I traced its tracks through some mud, where it tugged the rabbit to the brink of the nearest brook to be dipped and devoured.

"After the meal, Mr. Coon halted on an old bit of stump as gray as himself, close to where I lay under cover, trying to get a peep at his operations, but, unluckily, in my excitement I touched a bush, and broke a twig not as big as my little finger. I tell you he just jumped off that stump as if it scorched him, and disappeared."

"What about that tame coon you owned, Cy?" Dol asked. "You haven't got him now."

"Bless your heart, I should think not!" Here the student indulged in a chuckle of mirth. "That coon was the fun and bane of my life. No fear of my being dull while I had him! I had him as a present, when he was only a cub, from a man out here who is my special chum among woodsmen, Herb Heal, the guide in whose company we're going to explore for moose, and the soundest fellow in wind, limb, and temper that ever I had the luck to meet. I guess you English boys will say the same when you know him.

"Well! when my friend Herb bestowed upon me that baby raccoon, I called the little innocent 'Zip,' and kept him in-doors, letting him roam at will. But after he grew to manhood, I was obliged to banish him to our yard and chain him up; and there his piteous, sky-piercing calls, which seemed to come from the roof of a house near him, first showed me what a ventriloquist the animal can be."

"Why on earth did you banish him?" asked Neal.

"Because his plan of campaign, when loose, was to follow me about like a devoted cat, climbing over me whenever he got the chance, with slobbery fondness. But as soon as I was out of the way he'd steal every mortal thing I possessed, from my most precious instruments to my latest tie and handkerchiefs.

I never saw anything to equal his ingenuity in ferreting out such articles, and his incorrigible mischief in destroying them. I chained him in the yard after he had torn my father's silk hat into shreds, and made off with his favorite spectacles. Whether he wore them or not I don't know; he chewed up the case; the glasses no man thereafter saw. I couldn't endure his piteous cries for reconciliation while he was in banishment, so I gave him away to a friend who was suffering from an imaginary ailment, and needed rousing.

"Talking of fathers, boys, reminds me that I feel responsible to Francis Farrar, Esq., for the welfare of his lusty sons. Neal had a pretty tiring time last night, and only about two hours' sleep since. I don't suppose any of us are outrageously hungry, seeing that we had some kind of breakfast at an unearthly hour. Here we are at camp! I propose that we turn in, and try to sleep until noon. What do you say?"

Their leader having wound up his talk thus, neither of his comrades ventured to oppose his suggestion, though they felt little inclined for slumber.

"Pleasant day-dreams to you, fellows!" said Cyrus three minutes afterwards, flinging off his coat, and throwing himself on his mattress of boughs, while he wiped the steady drip of perspiration from his forehead and cheeks. "This day is going to be too warm for any more rushing. Our variable climate occasionally gives us these hot spells up to the middle of October; but they don't last. So much the better for us! We don't want sizzling days and oppressive nights, with mosquitoes and black flies to make us miserable. October in this country is the camper's ideal — month" —

The last sentence was broken by a great yawn, followed presently by a snort and an attempt at a shout, which quavered away into a queer little whine. Garst had passed into dreamland, where men revel in fragmentary memories and pell-mell visions.

## CHAPTER VI.

#### AFTER BLACK DUCKS.

IF Cyrus's dreams were ruffled after the morning's excitement, those of his comrades were a perfect chaos.

A slight wind hummed wordless songs through the tasselled tops of the pine-trees about the camp. The music was tender and drowsy as a mother's lullaby. Contrary to their expectations, Neal and Dol were lulled to sleep by it like babies, with a feeling as if some guardian spirit were gliding among the tree-tops.

But when slumber held them, when the murmur increased to a surge of sound, sank to a ripple and again rolled forth, in their dreams they imagined it the scurrying of a deer's hoofs along some lonely forest deer-path, the rustling of a buck through bushes, the splashing of a mighty moose among lily-pads and grasses at the margin of a dark pond, the startled cluck of a coon. In fact, that rolling music of the pines was translated into every forest sound which they had heard, or expected to hear.

The excitement of wild scenes, new sensations, strange knowledge, still thrilled them even in sleep. Their visions were accordingly wild, rushing, jumbled, yet all set in a light so bright as to be bewildering — a sign that health and happiness as great as human boys can enjoy were the possession of the dreamers.

By and by their pulses grew steadier. Out of this confused rush of imaginings grew in the mind of each one steady, absorbing dream. Neal fancied that he was on the top of Old Squaw Mountain, and that beneath, above, around him, sounded the strangely prolonged weird call, which he had heard at a distance on the previous night while Cyrus was recovering the jack-light. Owing to the ever-changing excitements of camp-life, he had not questioned his comrade again about it.

Dol's visions resolved themselves into a

mighty coon hunt. He tossed on his pine boughs, kicked and jabbered in his sleep, with sundry odd little cries and untranslatable mutterings, —

"Go it, Tiger! Go it, old dog! There he is — up the tree! Ah" (disgustedly), "you're no good!"

A lull. Then the dreamer rolled out a string of what may be called gibberish, seeing that it consisted of fragments of words and was unintelligible, followed by, —

"The coon's eating the pork — no, he's b-b-b-barking it! Hu-loo-oo!"

"Oh, say, Chick, give us a chance! We can't sleep with you chirping into our ears."

It was Cyrus who spoke, shaking with drowsy laughter, and Cyrus's big hand gently shook the dreamer's arm.

"What? what? wh-wh-at?" gasped Dol, awaking. "I wasn't talking out loud, was I?"

"Not talking aloud! Well, I should smile!" answered the camp captain. "You were making as much noise as a loon, and that's the noisiest thing I know. Go to sleep again, young one, and don't have any more crazy spells before dinner-time."

Cyrus removed his hand, shut his eyes, and in a minute or two was breathing heavily. Neal, who had been aroused too, followed his example, laughing and mumbling something about " it's being an old trick of Dol's to hunt in his sleep."

But the junior member of the party remained awake. After his dreams had been dissipated he cared no more for slumber. When he could venture it without disturbing his companions, he rose to a sitting posture, and, after squatting for a while in meditation, got on his feet, picked up his coat and moccasins, and, stealthily as an Indian, crept out of the hut.

The rolling music among the pine-tops had died down; only at long intervals a soft, random rustle swept through them. It was nearly midday. The camp-fire was almost dead, quenched by the dazzling sunlight which fell in patches on the camping-ground, and flooded the clearing beyond the shadow of the pines.

Moreover, the camping-ground was deserted. Neither Uncle Eb nor Tiger could be seen, though Dol's eyes sought for them wistfully. But something caught his attention. It was a ray of light filtering through the pine boughs and glinting on the trigger of an old-fashioned muzzle-loading shot-gun, which leaned against a corner of the hut. An ancient, glistening powder-horn and a coon-skin ammunition pouch hung above it.

Dol lifted the antiquated weapon, withdrew to a short distance, and examined it closely. He knew it belonged to the guide, but was rarely used by him since he had purchased the 44-calibre Winchester rifle, with which he could do uncommon feats in shooting.

The shot-gun interested the boy mightily. There was a facsimile of it, swathed in green baize, stowed away somewhere in his father's house in Manchester. The first time he had ever used fire-arms was on a memorable day when his fingers pulled its trigger in his father's garden under Neal's direction, and a lean starling fell before his shot. After that he had often taken out a fowling-piece of a newer style, and had done pretty well with it too.

As he handled the shot-gun, which the guide had bought away back in the year '55, musing about it under the pines, the thought suddenly tumbled out of a corner of his brain that at present there was a brilliant opportunity for him to use the gun and all the shooting skill he possessed for the benefit of his comrades and himself.

There was no meat in the camp for dinner or supper save the pork on which they had feasted since they arrived there, and that was fast giving out. Cyrus, in addition to his knapsack, had hauled over from Greenville, where articles of camp fare could be procured in abundance, a goodly supply of tea, coffee, condensed milk, flour, salt, sugar, etc., in a stout canvas bag, Neal at intervals helping him with the burden. For the rest he had trusted to Nature's larder, and such food as he might purchase from his guides, desiring to go into the woods as "light" as possible.

Uncle Eb had baked bread for his guests after a fashion of his own on the camp frying-pan, setting the pan on some glowing coals a foot or so from the fire ; he had fried unlimited flapjacks, and had cheerfully placed what stores he had at their disposal. His three luxuries were novelties to the English lads, being pork, maple sugar, — drawn from the beautiful maple-trees near his camp, — and a small wooden keg of sticky, dark molasses. The sugar was the only one which Dol found palatable ; and he knew that the Bostonian, Cyrus, shared his feeling. To tell the truth, the juvenile Adolphus was not fastidious, but he was suddenly seized with an ambitious desire to vary the diet of the camp.

"Uncle Eb said that I could use this 'ole fuzzee,' as he called it, whenever I liked," he muttered, looking wistfully at the shot-gun; "and I've a big mind to give those lazy fellows in there a surprise. They spent the night out jacking, and didn't get any meat because Cyrus let Neal do the shooting, and he bungled it. It's my turn next to go after deer, but I'm not going to wait for that."

Here his steel-gray eyes fell on the moccasins which he had not yet put on, and struck fire instantly. His ambition was doubled. For if there is one thing more than another which in the forest will stir the pluck of a novice, and make him feel like an old woodsman, it is the sight of his Indian footwear. Dol put his on, admired their light, comfortable feeling, their soft buckskin, and rashly decided that he could dispense with the loose inner soles which Cyrus had fitted into them to protect his feet.

Then, being very much of a stranger to American woods, he communed with himself after this fashion, —

"Cyrus says that different tribes of Indians wear differently made moccasins, and one redskin, if he sees the tracks of another in soft mud or snow, can tell what tribe he belongs to by his footmarks. That's funny! I suppose if any old brave was knocking about and saw my tracks in a boggy spot, he'd think it was a Kickapoo who had passed that way — not Dol Farrar of Manchester, England. These are of the shape worn by the Kickapoo tribe — so Cy says.

"I'm the kid of the camp, I know," he went on, with another flash in his eyes, as if there was a bit of flint somewhere in his make-up which had struck their steel. But I'll be bound I can do as well or better than the others can. I'm off now to Squaw Pond. I think I can follow the trail easily enough. Uncle Eb showed me yesterday where he had spotted some of the trees all the way along to the water. And if I don't shoot a couple of black ducks for dinner or supper, I'm a duffer, and not fit for camping."

He took down the powder-horn and slung it round him, saw that there was plenty of meat in the ragged coon-skin ammunition pouch which hung beside it, fastened that to his belt, slipped on his coat, and started off, with the "ole fuzzee" on his shoulder.

Never a sound did he make as he crossed

the clearing, passing the clump of bushes behind which Cyrus and Neal had lingered on the previous night to hear Uncle Eb's song. Owing to his Indian footwear, silently as the gliding redskin himself he entered the woods at a point where he saw a tree with a fresh notch carved in it. He knew this marked the beginning of the "blazed trail," and that he must be very wide-awake and show considerable "gumption" if he wanted to follow that line to the pond.

Not every tree was spotted. Only at intervals of fifteen or twenty yards he came upon a trunk with two small pieces chopped out of it on opposite sides. These were Uncle Eb's way-marks. One set of notches would catch his eye as he went towards the water, the other would lead him back to camp. Once or twice Dol got away from the trail, but he quickly found it again; and in due time emerged from the forest twilight into the broad glare of the sun, to see Squaw Pond lying before him like a miniature mother-of-pearl sea, so protected by its evergreen woods that scarcely a ripple stirred it.

He heard the shrill, wild call of a loon, the noisy bird to which Cyrus had likened him, and saw its white breast rising above the water, as it swam about among the reeds near the opposite bank. The cry was oft repeated, making an unearthly din, now joyous, now dreary, among the echoes around the lake.

Dol paused for a minute to listen; but he was bent on business, and did not want to be very long away from camp lest his absence should cause alarm. He took a careful survey of the scene. Not beholding any fleet of black ducks as yet, he loaded his gun, and warily proceeded along the bank towards the head of the pond.

Keeping a sharp lookout, he by and by detected something moving among the water grasses a little way ahead, and heard a hoarse, squalling "Quack! quack!"

Immediately afterwards a flock of half a dozen ducks sailed forth from their shelter, nodding and quacking inquisitively.

A wild drumming was at Dol's heart, and a reckless singing in his ears, as he raised his gun to his shoulder, and fired among them. Nevertheless, his aim was sure and deadly. Two quackers were killed with one shot! The others rose from the water, and with much fluttering and hoarse noise winged their way to safety.

"How'll they be for meat, I wonder? Won't I have a crow over those fellows?" shouted Adolphus aloud, with a yell entirely worthy of a Kickapoo Indian, when he had recovered from surprise at the success of his own shot.

He laid down the gun, pulled off his moccasins and socks, rolled up his trousers, and waded in for the prize. Truly luck was with him — so far — in his first venture in this region of the unknown. The water was so shallow that, having grabbed the ducks, he splashed out of it, kicking shiny drops from his toes, without wetting an inch of his garments.

"I'm the kid of the camp, I know; but I'll be the first fellow to bring any decent meat into it. Hooray!" he whooped again. "Shouldn't wonder if these moccasins brought me wonderful luck; one can steal about so quietly in them."

He had hit upon the supreme advantage which the Indian footwear possesses over every other for the woodsman. A little later he was to learn its disadvantage, having, with foreign inexperience, disdained the extra soles because they were not "Indian" enough for his taste; for the soft buckskin could not protect from roots and stones a wearer whose flesh was not hardened to every kind of forest travelling.

But at present Dol bepraised his moccasins; for they had enabled him to sneak upon his birds, the wildest of the duck tribe, who generally, at a single hoarse "Quack!" from their leader, will cease their antics in lake or stream, and disappear like a skimming breeze before a sportsman can get a fair shot at them.

For a quarter of an hour Dol Farrar sat by this forest pond engaged in the cheerful occupation of "booming himself," as his friend Cyrus would have said. He told himself that he had made a pretty smart beginning, not alone in shooting a brace of black ducks, but in successfully following a difficult trail on his fourth day in the woods. Henceforth, he thought, there would be little reason for him to dread the unknown in this great wilderness.

He reclothed his legs, gathered the stiffening claws of the defunct quackers in his left hand, picked up his empty "ole fuzzee," which had done such good service despite its age, and set forth on his return to camp.

Retracing his steps along the bank, after some searching he found the beginning of the trail, and started along it with a know-it-all, cheerful confidence in the little bit of woodlore which he had acquired. Hence he now found it considerably more difficult to follow the spotted trees. His brain was excited and preoccupied; and when once in fancied security he suffered his eyes and thoughts to stray for a minute from the trail, every unfamiliar woodland sight and sound tempted them to wander farther.

First it was an old fox, which poked its sharp, inquisitive nose out of a patch of undergrowth near at hand. Dol uttered a mad "Whoop-ee!" and heedlessly dashed off a few steps in pursuit. Reynard whisked his brush as much as to say, "You can't get the better of me, stranger!" and defiantly trotted away.

Recovering his senses, the boy managed to recover the trail too, and was keeping to it carefully when a second temptation beset him. A chattering squirrel, seated on the low bough of a maple-tree, with his fore paws against his white breast, his eyes like twinkling beads, and his restless little head playing bo-peep with the intruding boy, began to scold the latter for venturing into his forest playground.

Dol's first thought was full of delighted interest. His second was a sanguinary one; namely, that a pair of ducks would only be one meal for four campers who were "camp-hungry," and that Uncle Eb had spoken of squirrels as "fust-rate eatin'." He handled his gun uncertainly, deliberating whether or not he would load it, and try a shot at the bright-eyed chatterbox.

Before he had decided one way or the other, the squirrel, still scolding and playing bo-peep, scampered off his bough, and up the trunk of the maple. Thence he quickly made good his escape from one tree to another, affording a whisking, momentary view now and again of his white breast or bushy tail. Dol absolutely forgot the blazed trail, forgot the stories which he had heard about forest perils, forgot every earthly thing but his admiration for the pretty, tantalizing fellow; though to do the lad justice, he soon came to the conclusion that the camp must be in a worse strait for want of provisions before he could have the heart to shoot him. He gave chase nevertheless, plunging along in a zig-zag way over a carpet of moss and dry pine-needles, and through some dense tangles of undergrowth, uttering a welcoming screech whenever he saw the bright eyes of the little trickster peering down at him from a bough.

He had travelled farther than he knew before his interest in the game waned. He began to feel that it was rather beneath the dignity of a fellow who wore moccasins, carried coon-skin pouch and powder-horn, and who was bound for remote solitudes in search of the lordly moose, to be interested in such an insignificant phase of forest life as the doings of a red squirrel.

Then he started back to find the trail. He walked a considerable distance. He searched hither and thither, straining his eyes anxiously through the bewildering gloom of the forest, but never a notched tree could he see. Whereupon Dol Farrar called himself some pretty hard names. He remarked that he had been a "hair-brained fool" and a "greenhorn" ever to leave the spotted track, but that he wasn't going to be "downed;" he would search until he found it.

And he certainly was enough of a greenhorn not to know that every step he now took was carrying him away from the trail, and plunging him into a hopeless, pathless labyrinth of woods. For Dol had lost all knowledge of directions, and was completely "turned round;" which means that he was miserably lost.

The disaster came about in this way. The forest here was very dense, the giant trees interlocked above his head letting so little light filter through their foliage that he could scarcely see twenty yards ahead of him, and that in a puzzling, shadowy gloom resembling an English twilight.

When he ceased chasing the squirrel, he imagined that he retraced his steps directly towards the point where he had quitted the trail. In reality, seeing nothing to aim for in this bewildering maze of endless trees, turned out of his way continually as he dodged in and out around massive trunks, he gradually worked farther and farther off the course by which he had come, drifting in random directions like a rudderless ship on mid-ocean. This helpless state is called, in the phraseology of the northern woods, being "turned round."

But Dol Farrar was spared for the present a thorough realization of the dreadful mishap which had befallen him. He had a shocked, breathless, flurried feeling, as if scales had suddenly fallen from his eyes, and he saw the dangers of the unknown as he had not before seen them. But even in the midst of abusing himself for his rash self-confidence, he uttered a cheerful " Hurrah ! "

"Why, good gracious!" he cried. "Here's another trail! Now, where on earth does this lead to? I don't see any spotted trees" — looking carefully about — "but it's a well-beaten track, a regular plain path, where people have been walking. It must lead to our camp. I'll follow it up, anyhow. That will be better than dodging around here until I get 'wheels in my head,' as Uncle Eb says he did once when he lost his way in the woods, and kept wandering round and round in a circle."

Puffing with excitement and revived hope, the boy started off on this new trail, which he blessed at first — oh, how he blessed it! — as if it had been a golden clew to lead him out of his difficulty. To be sure, it was not a blazed trail; there were no notches in the trees, but the ground showed distinct signs of being frequently and recently travelled over. Though footprints were not traceable, moss, earth, and in some places the forest undergrowth of dwarfed bushes, were thoroughly pressed and trodden.

Dol never doubted but that it was a human

trail, a track continually used by some woodsman; but he thought that the unknown traveller, whoever he was, must have agile legs and a taste for athletics, for many times he had to hoist himself, his gun, and the ducks over some big windfall which lay right across the way. The dead quackers he pitched before him, fearing that by the time he got back to camp — if ever he did? — their flesh would be too bruised to look like respectable meat; for he was obliged to have one hand free to help him in scrambling over each fallen tree.

Once or twice this strange trail led him through thickets where the bushes grew so high as to lash his face. He came to regard slippery, projecting roots and rough stones, which galled his feet, protected only by the thin soles of his moccasins, as matters of course. His wind decreased, and his blessings ceased. Yet he followed on, walking, walking, interminably walking, with now and again an interval of climbing or stumbling headlong, accompanied by ejaculations of thankfulness that his gun was not loaded.

His breath came in hot, strangling gasps, the veins in his head were swollen and stinging like whipcords, there was a dull, pounding noise in his ears, and a drumming at his heart. He confessed that he was thoroughly "winded" when he had been following the trail for nearly two hours, so he seated himself upon a withered stump beside it to rest.

He had relinquished the idea that the track would bring him out near Uncle Eb's camp. Had it led thither, he would have rejoined his comrades long before this. His only hope now was that by patiently following it on he might reach the camp of some other traveller, or the lonely log cabin of a pioneer farmer. He had heard of such farm-settlements being scattered here and there on forest clearings.

So presently Dol Farrar got to his feet again, when he had recovered breath and strength, and told himself pluckily that "he wasn't going to knock under," that "he had been in bad scrapes before now, and had not shown the white feather." He gritted his teeth, and resolved that he would not show that craven pinion, even in the desperate solitude of these baffling woods where no eye could see his weakness. He did not want to have a secret, humiliating memory by and by that he had been faltering and distracted when his life depended on his wits and endurance.

He squared his shoulders sturdily, as if to

make the most of the budding manhood that was in him, and trudged ahead. And, indeed, he had need to take his courage in both hands, and force it to stand by him; for he had not gone far when, though the forest still continued dense, he became aware that he was beginning a steep ascent. Was the trail going to lead him up a mountain-side? The way grew yet more rugged. Every step was a misery. Jagged edges of rock and never-ending roots seemed to brand themselves with burning friction upon his feet, through their soft buckskin covering. He tried to hearten himself into a belief that he must soon reach some mountain camp or settlement.

But a bleak horror threw a gray shade upon his face as his staring eyes saw that the trail was growing fainter — fainter — fainter. At the foot of a steep crag, where a mass of earth, stones, and dead spruce-trees showed that there had lately been a landslide on the mountain above, he lost it altogether. It had led him to a pile of rubbish.

## CHAPTER VII.

#### A FOREST GUIDE-POST.

AT the foot of that crag Dol stood still, while a great shiver crept from his neck up the back of his head, stirring his hair. He peered in every direction; but there was no sign of a camp, nothing to show that any human foot before his had disturbed the solitude of this mountain-side, and no further marks on the ground, save one impression on a bed of earth at his feet where some animal had lately lain.

The disappointment was stupefying.

At last a fog of terror settled down upon him,—a fog which blotted out every sight and sound, blotted out even his own thoughts, all except one, which, like a danger-signal in a mist, kept booming through his brain: "Lost! Lost!"

By and by he was sitting on the piled-up stones and dirt of the slide; but he had no remembrance of getting to this resting-place, for he was still befogged.

Something snorted close to his right ear, — a loud snort, which banished stupor, and set his pulses jumping. It was a deer, a beautiful doe in a coat of reddish-drab, matching the autumnal tints of the forest, wherever maples, birches, and cedars mingled with the evergreens. She had bounded upon him suddenly from behind a dead spruce and a mound of earth.

It was long since the game on this part of the mountain had been disturbed. Madam Doe had in all probability never seen a man before, therefore her behavior was not peculiar. A shock of surprise thrilled through her graceful body as she vented that snort, when she caught sight of the new-fangled gray animal who had intruded upon her world, and who sat spell-bound, gazing at her with hopeless eyes, in which gradually a light broke.

But she did not fear him, — this creature in gray. She stood stock-still, and stared at him, so near that he could see her wink her starry eyes, with the white rings round them. She stamped one hoof, kicked an insect from her ear with another, snorted again, wheeled around, and at last broke away for the thick shelter of the trees, lightly and swiftly as a breeze which skims from one thicket to another.

Seeing his mother go for the woods, her spotted fawn, which had been frolicking among the branches of the fallen spruce-tree, skipped from it, passed Dol with a bound which carried him a few feet, and disappeared like a whiff too.

Here was a rouser, indeed, which no boy, unless he was in a far-gone state of suffering, could withstand. Dol Farrar forgot his terrible predicament. The fog had cleared away from his senses, leaving him free to think and act once more.

"Well, I never!" he ejaculated, springing to his feet in amazement. "Wasn't she a beauty? And wasn't she a snorter? I didn't think a deer could make such a row as that. And to stand still and stare at me! I wonder whether she took me for some new-fashioned sort of animal or a gray old stump."

It was a few minutes before he again thought of his plight, and then he was not overcome. He stood perfectly still, trying to review the position coolly, and to get a tight grip of his feelings, so that terror might not again master him.

"I'm in a worse scrape than I ever dreamt of," he muttered, puckering his forehead to do some tall thinking. "And I must do something to get out of it. But what? That's the question.

"I wonder if I loaded this 'ole fuzzee,'"— the lad was making a valiant effort to cheer himself by being jocular, — " and blazed away with it for a while like mad, whether there is any human being around who would hear me. Some fellow might be hunting or trapping in this part of the forest, or farther up the mountain. But what a blockhead I am! Why on earth didn't I do that before I started on this wretched trail?"

But alas! as this was Dol Farrar's first adventure in American woods, it had not occurred to him to do the right thing at the right time. Had he fired a round of signal shots when first he lost the line of spotted trees, he would probably have been heard at his camp, and would have been spared the worst scare he ever had in his life. The negligence was scarcely his fault, however; for

Cyrus Garst, who had never before undertaken the responsibility of entertaining a pair of inexperienced boys in woodland quarters, had not, at this early stage of the trip, arranged with his comrades to fire a certain number of shots to signify " Help wanted ! " if one of them should stray, or otherwise get into trouble. The idea now cropped up in Dol's perplexed mind, through a confused recollection of tales about forest misadventures which Uncle Eb had told him by the cheery camp-fire.

So he loaded the old shot-gun. It belched forth fire and smoke into space. And the thunder of his shot went rolling off in a reverberating din among the mountain echoes, until a hundred tongues repeated his appeal for help. Again he loaded rapidly and fired. And yet again, with nervous, eager fingers. So on, till he had let off half a dozen shots in quick succession.

Then he waited, listening as if every pulse in his body had suddenly become an ear.

But when the last growling echo had died away, not a sound broke the almost absolute silence on the mountain-side. Evidently not a human soul was near enough to hear or understand his signals of distress.

In these bitter minutes some sensations ran through Dol Farrar which he had never known before; and, as he afterwards expressed it, "they were enough to cover any fellow with goose-flesh."

He felt that he had reached the dreariest point of the unknown, and was a lonely, drifting atom in this immense solitude of forest and rock.

Never in his life before or afterwards did he come so near to Point Despair as when he stumbled down the mountain, spurning that treacherous trail, and going wherever his jaded feet found travelling tolerably easy. He had picked up the shot-gun; but the black ducks, the primary cause of his misadventure, he clean forgot, leaving them lying amid the chaos at the foot of the crag, to have their bones picked by some lucky raccoon or fox.

Wandering along in a zigzag way, he by and by reached the base of the mountain at a point where there was a break in the forest. A patch of dreary-looking swamp was before him, covered with clumps of alder-bushes — a true Slough of Despond.

Dol Farrar knew none of the miseries of plunging through an alder-swamp, but he luckily recalled in time a warning from Cyrus that a slight wetting would render his moccasins useless. While he halted undecidedly on its brink, he pulled out his watch; one glance at this, and another at the sky, which now lay open like a scroll above him, gave him a sickening shock. He had started from camp at noon; now it was after five o'clock. Little more than another hour, and not twilight, but the blackness of a total eclipse, would reign in the forest.

The blood rushed to his head, and his mouth grew feverish at the thought. As he licked his cracking lips, he caught a faint, tinkling, rumbling sound of falling water somewhere to the right. Of a sudden his sufferings of mind and body were merged into one burning desire to drink, and he turned eagerly in that direction.

At the edge of the woods he found a little fairy, foamy waterfall, which had tumbled down from the mountain to be lost in the dismal swamp. But Dol felt that it had accomplished its mission when he unfastened the tin drinking-mug which hung from his belt, and drank — drank — drank! He straightened himself again, feeling that some of the bubbling life of the mountain torrent had passed into him. His eyes lit on a towering pine-tree just beyond it. And then —

Well! if that sky-piercing pine had suddenly changed at a jump into a gray post, bearing the inscription, "One mile to Boston," Dol Farrar could not have been more astonished and relieved than when he saw for the first time a rude forest guide-post.

To the dark, knotted trunk was fastened a piece of light, delicate bark, stripped from a white-birch tree. On this was scrawled in big letters, by some instrument evidently not intended for penmanship : —

"FOLLOW THE BLAZED TRAIL AND YOU ARE SAFE."

"Another blazed trail! Hurrah!" shouted Dol. "Won't I follow it? I never will follow any other again if I live to be a hundred, and come to these woods every year till I die!"

The height of his relief could only be measured by the depth of his past misery, which would truly have been enough to set a weaker boy crazy. With watering eyes and panting breaths that came near to being sobs of gladness, he started upon the new trail. It led him off into the forest surrounding the swamp.

The pine that had been chosen for guide-post was the first in the line of spotted trees. The others followed it closely, with intervals of eight or ten yards between them; and as the notches in their trunks were freshly cut, Dol followed the track without any difficulty for twenty minutes. He had a suspicion that he was nearing the end of it; though he was still in forest gloom, with light coming in meagre, ever-lessening streaks through the pine-tufts above. Then he started more violently than when the deer snorted near his ear.

Suddenly and shrilly the blast of a horn rang through the darkening woodland aisles, followed, after a pause of a minute or two, by a second and louder blast.

Then a well-pitched, far-reaching voice sang out: —

"Come to supper, boys! Come to supper!"

"Good gracious!" said Dol, conscious on the instant that he was as hollow as a drum. "There are enough surprises in these forests to raise the hair on a fellow's head half a dozen times a day!"

A matter of forty yards more, and a burst of light swam before his eyes. He had reached the end of the blazed trail.

## CHAPTER VIII.

#### ANOTHER CAMP.

"HELLO! Come to supper, boys! Come to supper right away!"

Half eagerly, half shrinkingly, Dol emerged from the woods, feeling a very torment of hunger quickened in him by the tantalizing sound of that oft-repeated invitation.

A sight met him which, because of what went before and all that came after, will be forever chief among the forest pictures which rise in exciting panorama before his memory, when camping is a thing of the past.

A broad dash of evening light, the sun's afterglow, fell upon a patch of clearing bordered by clumps of slim, outstanding pines, the scouts of their massive brethren. That this was used as a camping-ground the first glance revealed. A camp which looked to the tired eyes of the lost boy a real "home-camp," though it consisted of rude log cabins, occupied it. A couple of birch-bark canoes reposed amid a network of projecting roots. Withered stumps and tree-tops littered the ground.

In the foreground of the picture stood a man with a horn in his uplifted hand, which he had just taken from his mouth. He was minus a coat; and the rough-and-tumble disarray of his attire showed that he had been lounging by his camp-fire, or perhaps overseeing the preparation of supper. Dol had a vague impression that the individual was not a forest-guide like Uncle Eb, nor a rough lumberman such as he had heard of. He would have taken him for a pioneer farmer, — not having yet encountered such a character, — but there could be no farm on this little bit of clearing. And he was too dazed to see that there were signs of a cultivated intelligence in the tanned, beaming face under the horn-blower's broad-brimmed hat. Indeed, the hat itself, its wearer, log huts, canoes, and trees seemed to have a strange propensity to waltz before the lad's eyes, and there was a queer waving sensation in his own legs, as if they, too, would join in the spinning movement. For as he advanced into the light out of the sombre shadows, a dizziness from long tramping in the woods, and from a hunger such as he had never before experienced, overcame him. He reeled against an outstanding tree, troubled by an affliction which Uncle Eb had called "wheels in his head."

"Ho! you boys. Where in thunder are you? Come to supper, or the venison will be spoiled!" shouted the possessor of the horn again, shutting one eye into which a crimson ray was pouring, while he swept the skirts of the woods with the other; and there was music as well as bluster in his shout.

Lo! the first to answer this fetching invitation was the foot-sore, leg-weary boy, pale from exhaustion, with his strange equipment of powder-horn, coonskin pouch, and ancient shot-gun, who, getting partly the better of his giddiness, crossed the clearing slowly, as if he was groping his way. Within a few feet of the horn-blower he halted; for the man had lowered his horn, and was gazing at him with keen, questioning eyes. Dol tried to find suitable speech to express his need; but though words came with considerable effort, his voice sounded hoarse and creaky in his own ears, and threatened to crack off altogether.

He was doing his best to brace up and speak plainly, when his sentence was stopped by a noise of pounding footsteps. The next moment he saw himself surrounded by three well-grown, daring-looking lads, one about his own age, one older, one younger, who were gazing at him with critical curiosity. All the pluck in Dol Farrar rose to meet this emergency. He felt as if his legs were threatening to smash under him like pipe-stems. There was a whirling and buzzing in his head. It seemed as if his words had such a long way to travel from his brain to his tongue that they got confused and changed before he uttered them.

But through it all he was conscious of one clear thought: that he was an Old-World boy on parade before these strapping New-World lads. He set his teeth, drove his gun hard against the ground, and, as it were, anchored himself to it, while strange, doubting lights came into his eyes as he tried to get a grip of his senses.

He succeeded. At last he addressed the

DOL SIGHTS A FRIENDLY CAMP.

gentleman with the horn, knowing that he was speaking to the point, —

"Good-evening, sir," he said. "I — I — we're camping out somewhere in the woods. I — I got lost to-day. I've walked an awful distance. Perhaps you could tell me " —

But the man stepped suddenly forward, with a blaze of welcome in his eyes; for he saw the brave effort which the lad was making, and that his strength was giving out. He put a kindly arm through Dol's, as if to warmly greet a fellow-camper, but really to support him.

"I'll not tell you about anything until you've had a good, square meal," he said. "That's our way in woodland quarters, — to eat first, and talk afterwards. If you're lost, you've struck a friend's camp, and at the right time too, son; so cheer up! After supper you can tell us your yarn, and I guess we can set you right."

Here at last was a surprise of unmixed blessedness for poor Dol; namely, the brotherly hospitality which is always extended to a stranger in a Maine camp, whether that be the temporary home of a millionnaire or the shanty of a poor logger.

His new friend led him into the largest of

the cabins, which contained a fireplace built of huge stones, where red flames frisked around fragrant birch logs, a camp-bed of evergreen boughs about ten feet wide, a rude table, a bench, and a few stools of pine-wood.

Over the camp-fire was stooping a bright-eyed, muscular fellow, whose dress somewhat resembled Uncle Eb's, but who had no negro blood in his veins. He was frying meat; and such tempting whiffs mingled with the steam which floated up from his pan, that Dol's nostrils twitched, and his hungry longing grew almost unbearable as he inhaled them.

"I guess this chunk of ven'zon is about cooked, Doc," said this personage, as Dol's kindly host entered the hut, with him in tow, followed closely by the boys of his own camp.

"All right, then! Let's have it!" was the reply. "I'm pretty glad our camp-fare is decent to-night, Joe, for we've a visitor here; a hungry bird who has strayed from his own camp, and has wandered through the forest until he looks like a death's head. But we'll soon fix him up; won't we, Joe? Give him a mug of hot tea right away. Hot tea is worth a dozen of any other drink in the woods for a pick-me-up."

A spark of fun kindled in Dol's eyes when he heard himself described as "a hungry bird." It brightened into an appreciative beam as the reviving tea trickled down his throat.

"Eatin's wot he wants, I guess," said Joe, the camp guide and cook, placing some meat and a slab of bread of his own baking on a tin plate for the guest.

Dol began on them greedily; and though the first mouthful or two threatened to sicken him, his squeamishness wore off, and he gained strength with every morsel.

"How do you like Maine venison, my boy? Like it well enough to have another piece, eh?" asked his host, when he saw that the haggard, gray look was leaving the wanderer's face, and that the appalled, dazed expression, the result of being lost in the woods, had disappeared from his eyes.

"I think it's the best meat I ever tasted," answered Dol heartily. "It's so tender, and has a splendid taste."

"Ha! ha! It ought to be prime," chuckled the owner of the camp. "It was cut from the quarters of a buck which my nephew here, Royal Sinclair," pointing out the tallest of the three lads, "shot four days ago. He was a regular crackerjack — that buck! I mean, he was as fine a deer as ever I saw; weighed over two hundred pounds, had seven prongs to his horns on one side and six on the other. Royal is going to take the antlers home with him to Philadelphia. We were mighty glad to get him, too; for we have been camping here for five weeks, and were running short of provisions. Roy had quite an attack of buck-fever over it, though he didn't think he was killing the 'fatted calf, to entertain a visitor; did you, Roy?"

"I guess not, Uncle! But I'm pretty glad, all the same," answered Royal, with a smiling glance at Dol.

Young Farrar found himself in very pleasant quarters; and, now that he was recovering, his laugh rang from one log wall to the other.

"What's 'buck-fever'?" he questioned, while Joe filled his plate with more venison.

"A sort of disease of which you'll learn the meaning before you leave these woods," answered his host merrily. "It attacks a man when he's out after a deer, and makes him feel as if one leg stands firm under him, while the other shakes as if it had the palsy.

"Now I guess you'd like to know whose

camp you're in, my boy, and then you can tell your story. Well, to begin with the most useful member of the party. That knowing-looking fellow over there, who cooked your supper, is Joe Flint, the best guide that ever pulled a trigger or handled a frying-pan in this region — barring one. These three rascals," here the speaker beamed upon the strapping lads, with whom Dol had been exchanging sympathetic glances of curiosity, "are my nephews, Royal, Will, and Martin Sinclair. And I — I —

"Good gracious! Listen to that, Joe! What's up now? Another fellow lost in the woods? Somebody is firing a round with his rifle! Perhaps he wants help. Those are signal shots, anyhow!"

The camper whose horn had been Dol's signal of deliverance, broke off abruptly in his introductions, just as he had arrived at the most interesting point, and was proclaiming his own identity. He rattled off his short exclamations in excitement, and dashed out of the cabin, followed by Joe, his nephews, and Dol, the latter limping painfully, for his feet now felt like hot-water bags.

"That Winchester has spoken eight or ten times," said the leader, counting the shots fired by somebody away in the dark recesses of the forest from a powerful repeating-rifle. "Let's give the fellow, whoever he is, an answer, Joe!"

He seized his own rifle hastily, loaded the magazine with blank cartridges, and fired a noisy salute.

In the pause which followed, while all strained their ears to listen, the sound of a shrill, distant "Coo-hoo!" the woodsman's hail, reached them from the forest.

Joe instantly responded with a vehement "Coo-hoo! Coo-hoo-oo!" the first call being short and brisk, the second prolonged into a roar which showed the strength of the guide's lungs, — a roar that might carry for miles.

Shortly afterwards there was a crashing and tearing amid some undergrowth near the edge of the forest. A man bounded forth from the pitch-black shadows into the clearing, where a little daylight still lingered. As he approached the group, Dol, who was in the background, gave a startled, yearning cry; but it was drowned in a loud burst from his host.

"Why, Cyrus Garst!" exclaimed the latter, peering into the new-comer's face. "How goes it, man? I never expected to see you here. Surely you haven't come to grief in the woods? You look scared to death!"

Cyrus — for it was he — grasped the welcoming hand which the owner of this camp extended to him. But his dark eyes did not linger a moment meeting the other's. They turned hither and thither, flashing in all directions restlessly, like search-lights.

"I'm glad to see you, Doc," he said. "I didn't know you were anywhere near. But I'm half distracted just now. A youngster belonging to our camp is missing. I've been scouring the forest for hours, and firing signals, hoping he might hear them. But"—

Here Cyrus caught sight of Dol, who with a cry which in its changing inflections was longing, penitent, joyful, was making towards him. The Harvard student strode forward, and gripped the boy by his elbows. In the dusk their eyes were near together; Garst's were stern, Dol's blinking and unsteady.

"Adolphus Farrar," began Cyrus in a voice as if he was making an arrest, "have you been here in this camp, or where have you been, while your brother and I were searching the woods like maniacs? What unheard-of folly possessed you to go off by yourself?"

Dol made a gurgling attempt to answer,

but his voice rattled and died away in his throat. His eyes grew decidedly leaky.

"Say, Cyrus!" interrupted the man who had befriended him and now proved his champion, "let the youngster get breath and tell his story from start to finish before you blow him up. I guess he wasn't much to blame; and if he was, he has suffered for it. He found his way here not quite half an hour ago, so played out from wandering through the forest that he was ready to drop in his tracks. And I tell you he showed his grit too; for he managed to brace up and keep on his feet, though he was as exhausted a kid as ever I saw."

The "kid," forgiving this objectionable term because of the soothing allusion to a trying time when he had behaved like a man, winked and gulped to get rid of his emotion, and twisted his elbows out of Cyrus's hold. The latter lost his angry look, and released them.

"I must fire three shots to let Neal and Uncle Eb know I've found you," he said. "We parted company a while ago, and they're beating about the woods in another direction. Whoever first came upon any trace of you was to fire his rifle three times."

The signal was instantly given.

More far-reaching "Coo-hoos!" were exchanged. Ere long Neal was beside his brother, looking at him with eyes which showed the same tendency to leak that Dol's had done a while ago, and battling with a desire to squeeze the wanderer in a breathless hug. He relieved his feelings instead by "blowing up" Dol with withering fire and a rough choke in his voice.

But when, in response to an invitation from the genial camper whom Cyrus and Joe called "Doc," the whole party, guides included, had gathered around the camp-fire in the big log hut, and Dol told his story from start to finish, he became the hero of the evening.

His only fault had been a rash venturing into the unknown; and well it was that he had not followed the unknown to his death.

"Why, boy!" exclaimed Cyrus, with a strong shudder, when Dol had described the false trail which led him to the foot of the crag, "that wasn't a human trail at all. It was a deer-road. The deer spend their day up in the mountains, and come down to the ponds at evening to feed and drink. Now, a buck or doe in its regular journeys to and fro will follow one line, to which it becomes accustomed. Perhaps fifty others, seeing the ground trodden, will run in the same track. And there you have your well-used path, which looks as if it was made by men's feet!

"You may thank your lucky star, Dol, every hour of this night, that the false trail didn't lead you away — away — higher — higher — up the mountain, until you dropped in your tracks, and died there alone, as others have done before."

A shocked hush fell upon the group around the camp-fire. Even the guides were silent. But the fragrant birchen logs sputtered and glowed, darting out playful tongues of flame. They seemed to call upon everybody to dismiss gloomy thoughts of what might have been; to crack jokes, sing songs, tell yarns, and be as merry as befitted men who had a log hut for a shelter, fresh whiffs of forest air stealing to them through an open doorway, and such a camp-fire.

Joe began to prepare supper for the three who had searched so long and distractedly for Dol that they confessed to not having eaten for hours. While more venison was being cooked, the juveniles, American and English, who had been secretly taking stock of each other, cast aside restraint, and became as "chummy" as if they had been acquainted for years instead of hours.

Such a carnival of fun and noise was started through their combined efforts in the old log camp, that its owner declared he "couldn't hear himself think." Seizing his horn, he blew a blast which called for order.

"Say, my boy, let me have a look at your feet," he said, cornering Dol. "A deer-road isn't a king's highway, as I dare say you've found out to your cost. Pull off your moccasins and socks, and let me doctor your poor trotters."

Young Farrar very gladly did as he was bidden.

"Humph!" said his friend. "I thought so. They're a mass of bruises and blisters. You've been pretty well branded, son. Moccasins aren't much use to protect the feet from roots and sharp stones, if you happen to strike a bad place in forest travelling, unless you have taken the precaution to put double soles in them; didn't you know that? Now, Cyrus Garst," turning to the student, "you're all going to camp with us to-night. This lad can't tramp any more. As a doctor I forbid it."

"Are you a doctor, sir?" questioned Dol,

with a thrill of surprise, which he managed to conceal.

"Something of the kind, boy," answered his host, smiling. "I don't look much like a city physician, do I? I graduated from a medical college in Philadelphia, and took my degree. But I had an enthusiasm for the woods. One hour of forest life in dear old Maine was to me worth a year spent amid streets, alleys, and sky-scraping buildings; so I fixed my headquarters at Greenville, and have spent most of my time in the wilderness."

"Where every trapper, guide, and lumberman knows Dr. Phil Buck, whom they disrespectfully and affectionately call 'Doc,'" put in Cyrus. "And many a poor fellow owes his life or limbs to Doc's knowledge and nursing in some hard time of sickness, or after one of the dreadful accidents common in the forests."

Dol could well understand this; for he now was benefiting by Dr. Phil's lively desire to relieve suffering, and was silently breathing blessings on his head. The doctor had bathed his puffy feet in warm water taken from Joe's camp-kettle, and was anointing them with a healing salve, after which he tucked them into a loose pair of slippers of his own. Meanwhile, he chatted pleasantly.

"This isn't the first time that your friend Cyrus and I have run against each other in the wilds," he said, "nor the first time that we've camped together, either. Bless you! we could make you jump with some of our stories. Do you remember that night in '89, Cy, when you, with your guide, came upon me lying under a rough shelter of bark and spruce boughs, which I had rigged up for myself near Roaring Brook, on the side of Mount Katahdin?"

"I guess I do remember it," answered Cyrus, laughing.

"A mighty hungry man I was, too, that evening," went on Doc; "for I had no food left but one little package of soup-powder and a few beans. I had been trying all day to get a successful shot at a moose or deer, and muffed it every time. It wasn't the lucky side of the moon for me. Well, you behaved like the Good Samaritan to me, then, Cy; shared your meat and all your stuff, and we slept like twin brothers under my shelter."

"Yes; and a bear visited our temporary camp in the night!" exclaimed Cyrus, bursting into uproarious mirth over some overpoweringly funny recollection; "he made off with my knapsack, which I had left lying by the camp-fire. I suppose old Bruin thought he'd find something good in it to eat; but he didn't. So he tore my one extra shirt and every article in the pack to shreds, and chewed up the handle of my razor, so that I couldn't shave again until I got back to civilization, when I was as bristly as a porcupine."

"Perhaps Bruin tried to shave himself," suggested Dol.

"At all events, he had wisdom enough not to cut his throat," answered the story-teller. "We three — Doc, my guide, and myself — were stupidly tired, and slept so soundly that we did not discover the theft nor who the marauder was until the following morning. Then we found my knapsack gone, and the tracks of a huge bear in some soft earth near our shelter. We traced his footprints through a bog until we found the spot, not far off, where, overcome by greed or curiosity, he ripped up that strong leather knapsack as if it was *papier maché* and made hay of its contents."

The boys had all crowded near to listen. It was now the social hour for campers. By the camp-fire more reminiscences followed; and the two guides chimed in it with moose stories, bear stories, panther stories, wild tales of every imaginable and unimaginable kind of adventure, until the lads thought no mythology which they had ever learned could rival in marvels the forest lore.

At this opportune time, Neal suddenly thought of describing, or attempting to describe, that strangest of strange calls which he had heard, after the capsizing of the canoe, on the preceding night, when Cyrus and he were jacking for deer on Squaw Pond.

Joe grunted expressively. "So help me! it was the moose call!" he ejaculated. "What say, Doc?"

"I guess it was," answered Dr. Phil. "It was either the cow-moose herself calling, or some hunter imitating her with his birch-bark trumpet. It's a weird sort of experience, to hear that call for the first time; I shouldn't wonder if your heart went whack-whack, lad?"

"I only hope he'll get a chance to hear it again before he goes back to England," said Cyrus.

Forthwith, the Harvard man proceeded to explain that he was bent on pressing forward for a distance of sixty miles or so, to the heart of the wilderness, to search for moose, but that he intended to do the journey in a leisurely, zigzag fashion, camping for a couple of nights at various points, in order to do the honors of the forest to his English comrades.

"So you're English, are you! Ha! Ha! Ho! Ho!" exclaimed the doctor, looking at the young Farrars. "Well, I suppose we'll have to put our best foot foremost to give you a good time in American woods."

"I think that's what we're having, sir — such a jolly good time that we'll never forget it," answered Neal courteously.

"Yes, it's jolly enough now; but I tell you I didn't find it so to-day," grumbled Dol, while his eyes gleamed like polished steel with the light of present fun. "But as long as I live I'll remember the sound of your horn, Doctor, when I was dead-beat."

"Is that so? Well, I guess I'll have to make you a present of that horn, boy, when we part company, and you go back to civilization, and of the piece of birch-bark, too, which led you to our camp. 'Twas Joe who fixed that to the pine near the swamp; for my lads had a habit of following the trail to the alders, looking for moose or deer signs. He scrawled his sentence on it with the end of a cartridge. I guess it would be a sort of curiosity in England."

Dol whooped his delight.

"I'll put it under a glass shade! I'll "—

While he was casting about in his mind for some way of immortalizing that bit of white bark, Doc's genial bluster was heard again, —

"Come! come! you fellows! No more skylarking in this camp to-night! It's high time for all campers to be snoring. Turn in! Turn in!"

But nobody was in a hurry to obey the summons to bed. While hands and feet were being stretched out to the sizzling birch logs for a final toast, Royal Sinclair, who had a trick of speaking very quickly, with a slight click in his utterance, as if his tongue struck his teeth, began to pour some communications into Neal's ear in rapid dashes of talk, —

"This is just about the jolliest night we ever had in the forest, and we've had a staving time all through. We live in Philadelphia, and Uncle Phil — we call him 'Doc' like everybody else — brought us out here for our summer vacation. This old log camp was built several years ago by a hunting-party, of whom he was one. The walls were getting mouldy; but he cleaned up the largest of the huts, with Joe's help, and made it our headquarters. He never needs a guide himself; not a bit of it! He can find his way anywhere through the woods with his compass. But he is a good deal away, so he engaged Joe to go out with us.

"He often starts off at a moment's notice, and travels dozens of miles on foot, or in a birch canoe, if he hears of a bad accident far away in the forest. Sometimes a lumberman or trapper cuts his foot in two, or nearly chops off his leg with his axe; and these poor fellows would probably die while their comrades were lugging them through the woods on a litter, trying to reach a settlement, if it weren't for our Doc.

"Once in a while, when he comes to visit us in Philadelphia, a few people call him a crank, because he lives out here and dresses like a settler; but I call him a regular brick."

"So do I," said Neal with spirit.

"You're awfully lucky to be able to camp out during October," rattled on Roy. "That's the month for moose-hunting, jacking, and all the most exciting sort of fun. We have to go home in a day or two, for our school has reopened, unless"—

"When Royal Sinclair gets a streak of talking, you might as well try to bottle up the Mississippi as to stop him," said Dr. Phil, laughing. "I can't hear what he's saying, but I know that his tongue is clicking like a telegraph instrument. But I hope it has given its last message for to-night. You really must turn in, boys. I let you have an extra social hour, because to-morrow will be Sunday, a day of rest after the travels and excitements of the week. Think of it, lads! A Sunday in the woods — God's first cathedral! May it do us all good!"

The guide, Joe, built up the fire. Fresh birch logs blistered and sputtered as creeping curls of bluish flame enwrapped them. Kindling rapidly, they threw out fantastic lights, which danced like a regiment of red elves around the old log walls of the cabin.

"If a fellow could only drop off to sleep every night in the year seeing and smelling such a fire as that!" breathed Neal, as, accepting a share of Royal's blankets, he stretched his tired limbs on the evergreen mattress.

"Then life would be too jolly for anything," answered Roy.

## CHAPTER IX.

### A SUNDAY AMONG THE PINES.

"MEN and boys learn a good many wholesome lessons in the forest, one of which is that it pays better to take a day of rest in seven if they want to make the most of themselves and their opportunities. Therefore, lads, we'll do no tramping to-day. And we'll have a bit of a service by and by over there under the pines."

So spoke Doctor Phil on the following morning, when the two sets of campers, now one joyous, brotherly crowd, were sitting or lounging about the pine-wood table, leisurely emptying tin mugs of tea or coffee, and eating porridge and rolls of Joe's baking.

"You haven't told us yet, Cyrus," he went

on, "what point you're bound for. I know you're level-headed, and plan every forest trip beforehand, to economize time."

"Yes, a fellow likes to do that; it adds to the pleasures of anticipation," Garst answered. "But it's precious little use, after all, when you're visiting a region which is as full of surprises as an egg is full of meat. However, I have arranged to meet Herb Heal, the guide whom I generally employ, at a hunting-camp near Millinokett Lake."

"A good moose country," put in Doc.

"I know it. At all events, it is a good place for a home-camp; one can make excursions into the dense forests at the foot of Katahdin, which are unrivalled for big game — so Herb says, and he's an authority. These English fellows may expect to have an attack of buck-fever, or *moose-fever* rather, which will set their blood on fire. Not that we're out chiefly for killing; we're willing to let his mooseship keep a whole skin, and go in peace to replenish the forests, unless he grows cantankerous and charges us."

"If he happens to be an old bull, and gits his mad up, he may do that; it's as likely as not," chimed in Joe Flint, who was listening.

"Well, if there's a man in Maine who can be warranted to start a moose, and to follow up his trail until he gets a sight of him, living or dead, that man is Herb Heal," said the doctor. "And his adventures go ahead of those of any woodsman up to date. You must get him to tell you how he swam across a pond at the tail of a bull-moose, holding with his fingers and teeth to the creature's long hair, then got astraddle of its back, and severed its jugular vein with his hunting-knife. How's that! It was the liveliest swim I ever heard of. But I mustn't spoil his yarns. He must tell them himself.

"A fine son of the woods is Herb Heal!" went on the speaker, with enthusiasm. "I ran across him first five years ago, when he was trapping for fur-bearing animals in the dense forests you mentioned near the foot of Mount Katahdin. He had a partner with him then, a half-breed Indian, whom woodsmen called 'Cross-eyed Chris,' a willing, plucky, honest fellow when he was sober. But he loved fire-water. Let him once taste spirits, or smell them, and he went clean crazy. He did a dog's trick to Herb,— stole all his furs and savings, with a splendid pair of moose antlers, while he was away from camp one day, and skipped out of the State. Herb swore he'd shoot him. But I don't think he has ever come across him since. And if he should, he wouldn't stick to his threat. He's not built that way."

There was a general hum of interest over this story, which even Cyrus had not heard before.

"Now, how are you going to reach your camp on Millinokett Lake?" asked Dr. Phil, when the buzz had subsided. "That's the next question."

"We intend to tramp the entire distance by easy stages, and get there about the middle of October," answered young Garst for himself and his comrades. "Uncle Eb will go along with us as guide; and he'll supply a tent, so that we can rest for two or three nights at a time if we choose."

"Hum!" said the doctor doubtfully, laying his hand on Dol's shoulder. "This youngster oughtn't to do much tramping for a few days, Cyrus. That deer-road did up his feet pretty badly. I'll be travelling in your direction myself the day after to-morrow. I want to visit a farm-settlement within a dozen miles of the lake, where the farmer has a sickly child, the only treasure in his log shanty. The mite frets if Doc doesn't come to see her once in a while.

"Therefore, I propose that we join forces, and press forward together. I guess I'll keep my nephews out here for a week longer, and take the responsibility of their missing that time at school. Now that they have fallen in with your friends, it would be a shame to separate Young England and Young America without giving them a chance to get friendly."

Here Dr. Phil beamed upon the five boys, who, after one night in the forest, sleeping in a light-hearted row on the evergreen boughs, with their feet to the fire, had reached a brotherly intimacy which years of city life might not have bred.

"I further propose," he went on, "that we hire a roomy wagon and a pair of strong horses from a settler who has a clearing about two miles from here. There is an old logging-road which runs through the woods towards the point for which we're heading. We could follow that for the first half of our journey. It isn't a turnpike, you know. In fact, it's only a broad track where the underbrush has been cleared away, and the trees cut down, with strips of corduroy road sandwiched in. But the lumbermen still haul supplies over it to their camps, and I propose that we follow their example. We can pile our tent, camp duffle [stores], and all our packs into the wagon, together with the hero of the deer-road,"—winking at Dol,— "and the rest of us can take turns in riding. It will be a big lark for these youngsters to travel over a corduroy road. A very bracing ride they'll have in more senses than one; but they can spin plenty of yarns about it when they get home."

The "youngsters," one and all, signified their approval of the suggestion. Cyrus, who, as a college man, was above this category, was pleased to acquiesce too.

"When can we get the wagon, Doctor?" asked Neal, burning to press onward.

"Oh! the day after to-morrow, I guess. And now, lads!" Dr. Phil's voice was serious, but exultant, "we're a thoroughly happy set of fellows, in accord with each other and our surroundings. We feel our brains clear, our gladness springing up, and our lungs swelling to double their size with the whiffs which reach us from those sky-piercing pines yonder. So we will remember that 'the wide earth is our Father's temple.' Over there in the woods we will worship him, while millions of forest creatures about us, flying, bounding, or building, in obedience to his laws, simply worship too."

A music soft, deep, sighing, like the murmur of an organ under the fingers of a master musician, rolled through the pine-tops as the band of campers, guides included, followed Doc into the forest. They passed the clumps of slender trees near the camp, and reached a dimly-lit green aisle.

Towering pines, so tall and erect that they seemed shooting upward to kiss the clouds, were the pillars of their cathedral. Its roof of tasselled boughs was stabbed by flashing needles of sunlight, which let in a flickering, mellow radiance, and traced a pattern on the woodland carpet. Every whiff of forest air was natural incense.

Dr. Phil stood as if in the audience-chamber of the King, and removed his wide-brimmed hat.

"Now unto the King eternal, immortal, invisible, the only wise God, be honor and glory, for ever and ever. Amen!" he said.

Then Cyrus's voice led the worship.

"Praise God, from whom all blessings flow!"

he sang, in a strong, glad outburst.

Boys and guides, in a great chorus, swelled the familiar words. Each sweetly chirping woodland bird, after its own manner, echoed them. The music among the pine-tops mingled with them. The forest fairly rang with a magnificent, adoring Doxology.

"We ought to be decent kind of fellows after this," said Cyrus, when the little service was over.

And the doctor answered, —

"I tell you, boy, the church was never built where a man feels so ready to worship the God-Father in spirit and in truth as he does in the wild woods."

And looking on the six fresh, manly faces before him, Dr. Phil saw that this happy woodland trip would have grander results than adding to the campers' inches and to the breadth of their shoulders. For each one of them had realized this morning that behind all strength and beauties of forest growth, behind their own souls' gladness, was a Presence which they could "almost palpably feel."

## CHAPTER X.

### FORWARD ALL!

SPECULATIONS about the journey, and in especial about the corduroy road, were rife in the boys' minds during the forty and odd hours which elapsed between the Sunday service and the time of their start.

The travellers met at the settler's cabin early on Tuesday morning, having broken camp shortly after daybreak. On Monday evening Cyrus and Neal, with Uncle Eb, had returned to the bark hut to pack their knapsacks, and make ready for a forward march. On the way thither, it being just the hour for the deer to be running, — that is, descending from the hills for an evening meal, — Neal got a successful shot at a small two-year-old buck.

This was a stroke of luck for the campers, and a necessary deed of death. It supplied them with venison for their journey; and, as Cyrus said, "they had already put a shamefully big hole in Dr. Phil's stores, and must procure a respectable supply of meat to make up for it."

It also provided Tiger with plenty of bones to crunch during his master's absence; for the dog was left behind in charge of the hut, as indeed he often was for a week or more while Uncle Eb was away guiding. The sportsmen who engaged the latter's services were generally averse to the creature's presence with the party, lest he should scare their game.

Cyrus and Neal bade him a pathetic farewell, remembering the exciting fun he had given them with the raccoon. Dol sent him lots of approving messages, which were duly delivered, with rough pats and shakes, by Uncle Eb, who fully believed that the brute understood every word of them. Indeed, the sign language of Tiger's expressive tail confirmed this opinion.

Dol had remained at the log camp with his new friends, Dr. Phil thinking it well that he should rest his feet until the morning of the start. His brother promised to bring his knapsack and rifle to the settler's cabin. Uncle Eb repossessed himself of his shotgun, pouch, and powder-horn, which he carried back to his hut, and left under Tiger's protection, telling Dol that "if he wanted to bag any more black ducks he'd have to give 'em a dose wid de rifle, for he warn't a-goin' to lug dat ole fuzzee t'rough de woods."

It was the perfection of an October morning, sunshiny and pleasant, with a mellow freshness in the air which matched the mellow tints of the forest, when the travellers joined forces at the farm-settlement.

Engaged in the thrilling work of felling a pine-tree to extend his father's clearing, they found the settler's son, a brawny fellow about Cyrus's age, in buckskin leggings and coonskin cap, who wielded his axe with arms which were tough and knotted as pine limbs. He bawled to them in the forceful language of the backwoods, which to unaccustomed ears sounded a trifle barbaric, to keep out of the way until his tree had fallen.

When the pine at last tumbled earthward with a thud which reverberated for miles through the forest, he gave a mighty yell, waved his skin cap, and came towards the visitors.

"Hulloa, Lin!" boomed the doctor, greeting this native as an old acquaintance.

"Hello, Doc!" answered Lin. "By the great horn spoon! I didn't expect to see you here. Who are these fellers?"

The doctor introduced his comrades. Lin greeted them with bluff simplicity, and called them one and all by their Christian names as soon as these could be found out. Doc alone came in for his short title — if such it could be called. Luckily the campers of both nationalities, from Cyrus downward, were without any element of snobbery in their dispositions. It seemed to them only a jolly part of the untrammelled forest life that man should go back to his primitive relations with his brother man; that in the woods, as Doc said, "manhood should be the only passport," and that titles and distinctions should never be thought of by guides or anybody else. They were well-pleased to be taken simply for what they were, — jolly, companionable fellows, — and to be valued according to the amount of grit and good-temper they showed.

And they learned this morning to appreciate the pioneer courage and resolute spirit of the rugged settlers who had cleared a home for themselves amid the surrounding wilderness of forest and stream. Their roughness of speech was as nothing in comparison with their brave endurance of hardships, their deeds of heroism, and their free-handed hospitality.

Lin led his visitors straight to a log cabin, before which his father, a veteran woodsman, who bore the scars of bears' teeth upon his body, was digging and planting. This old farmer, too, greeted Doc as a friend, and when the wagon was talked about, was quite willing to do anything to serve him.

"But ye must have a square meal afore ye travel," he said. "Jerusha! I couldn't let ye go without eatin'. Mother!" shouting to his wife, who was inside the cabin. "Say, Mother! Ha'n't ye got somethin' fer these fellers to munch?"

Forthwith a big, rosy woman, who had herself fought a bear in her time, and had shot him, too, before he attacked her farmyard, hustled round, and got up such a meal as the travellers had not tasted since they entered the woods. They had a splendid "tuck-in," consisting of fried ham, boiled eggs, potatoes, hot bread, yellow butter, and coffee. And the meal was accompanied with thrilling stories from the lips of the old settler about the hardships and desperate scenes of earlier pioneering days. Doc coaxed him to relate these for the boys' benefit. And many eyes dilated as he told of blood-curdling adventures with the "lunk soos," or "Indian devil," the dreadful catamount or panther, which was once the terror of Maine woodsmen.

"So help me! I'd a heap sooner meet a ragin' lion than a panther," said the old man. "My own father came near to bein' eaten alive by one when I was a kid: He was workin' with a gang o' lumbermen in these forests at timber-makin', and was returnin' to their camp, when the beast bounced out of a thicket all of a suddint. Poor dad was skeered stiff. The thing screeched,— a screech so turrible that it was enough to turn a man's sweat to ice-water, an' a'most set him crazy. Dad hadn't no gun with him; so he shinned up the nighest tree like mad, an' hollered fit to bust his windpipe, hopin' t'other fellers at the camp 'ud hear him.

"But the panther made up another tree hard by, an' sprang 'pon him. Fust it grabbed dad by the heel. Then it tore a big piece out o' the calf of his leg, an' devoured it. Think of it, boys! Them's the sort o' dangers that the fust settlers an' lumbermen in these woods had to face.

"Wal, dad reckoned he was a goner, sure. But he managed to cut a limb from the tree with his huntin'-knife, an' tied the knife to the end of it. With that he fought the beast while his comrades, who had heard his mad yells, were gittin' to him. With the fust shot that one of 'em fired the catamount made off.

"Dad was the sickest man ye ever saw fer a spell. His wound healed after a bit, under the care of an Injun doctor; but his hair, which had been soot-black on that evenin' when he was returnin' to camp, was as white as milk afore he got about again; an' he was notional and narvous-like as long as he lived.

"He said the animal was like a tremenjous big cat, about four feet high an' five or six feet in length. It was a sort o' bluish-gray color. An' it had a very long tail curled up at the end, which it moved like a cat's.

"Boys, that catamount is the only animal that an Indian is skeered of. Ask a red man to hunt a moose, a bear, or a wolf, an' he's ready to follow it through forest an' swamp till he downs it or drops. But ask him to chase a panther, an' he'll shake his head an' say, 'He all one big debil!' He calls the beast, in his own lingo, 'lunk soos,' which means 'Injun devil;' an' so we woodsmen call it too."

It was at this moment that Lin put his head in at the cabin-door, and announced that "the wagon an' hosses war a' ready."

"Wal, boys, I swan! it's many a long year since a panther was seen in these forests, so ye needn't feel skeery about meetin' one," said the old settler, as he stood outside his log home, and watched his guests start. "I'll 'low ye won't find travellin' too easy 'long the ole corduroy road. Come again!"

There was much waving of hats as the wagon, a roomy, four-wheeled vehicle, moved off, with a creaking in its joints as if it were squealing a protest against its load, which consisted of the five lads, together with knapsacks, guns, tents, and the camp duffle.

"Forward, all!" shouted Dr. Phil, who had been chosen to act as captain of the two companies during the few days while they journeyed together.

Lin, who was charioteer, cracked a long whip above his horses. The boys cheered, while Doc, Cyrus, and the two guides fell behind, choosing to follow the wagon on foot for the first few miles of the journey.

"Where did you buy that, Lin?" asked

Neal, climbing over to a perch beside the driver, and pointing to a heavy Colt's revolver which the young settler was buckling round his waist.

"Didn't buy it. I traded a calf for it at Greenville more'n a year ago," was the reply. "Fust-rate gun it is, too, I vum! I've stood at our cabin-door, and killed many a buck with it. On'y 'tain't much good for tackling a bear. Wish't the bears ud get as scarce as the panthers! Then we'd be rid o' two master pests. Hello! Don't y'u git to tumbling out jist yet! That's on'y a circumstance to the jolts there'll be when we strike a bit o' corduroy road."

Lin Hathaway grabbed young Farrar by the elbow while he spoke, and held him steady with the horny hand which had swung the axe against the doomed pine-tree. For Neal had shown a sudden inclination to pitch headlong out of the wagon, as its right wheels were hoisted a foot or more above the left ones by rolling over a mossy bump in the ground.

For the first five miles the forest road had been simply constructed thus: First, the bushy undergrowth had been cut away and thrown to one side, the space cleared being about eight feet wide; then all trees growing in the range of this track had been sawn off close to the ground, and windfalls which barred the way were removed. It was a rude highway, with plenty of deformities, such as ends of rotting stumps, twisted roots, ridges and bumps which had never been levelled; yet it was beautiful beyond any smooth, well-graded road which the travellers had ever seen. As it wound along in graceful curves through the woods, it was shaded now by an emerald arch of evergreens, now by a royal crimson canopy of maple branches, while patches of buff, orange, and dull red commingled where other trees interlaced with these to whisper woodland secrets.

But the boys soon understood what Doc meant when he spoke of their having "a bracing ride in more senses than one;" for the motion of the wagon was a giddy series of jolts and bounces, with just sufficient interval between each shock for them to brace themselves, with stiffened backbones, for the next upheaval. They had already begun, as Royal said, " to have kinks in all their limbs," when Lin suddenly announced, —

"Yon's a bit o' corduroy road, I declar'!"

He pointed with his whip ahead, and the

travellers shot out their necks to see this novel highway. It extended for about a quarter of a mile over a swamp, and spoke volumes for the energy and ingenuity of the hardy lumbermen who constructed it.

These brawny heroes, who are fine types of American grit and manhood, when clearing a broad track over which their great timber logs could be hauled from the depths of the forest to the landing on some big river, had found the swampy tracts an impassable obstacle for animals trammelled with harness and a heavy load.

They bridged them by laying down logs cut to even lengths in a slightly slanting position across the way for the entire extent of miry ground. Each piece of timber was tightly wedged in by its fellow; nevertheless, there was a space of several inches between their rounded tops. Hence the track presented a striped appearance, which suggested to some spirited genius among woodsmen its name of "corduroy road."

"Well, Neal, do you think you can tell your folks a thing or two about forest travelling when you get back to England?" asked Doc, when the order of march was changed, young Farrar and the Sinclairs turning out to do their share of tramping, while the doctor, Cyrus, and the guides benefited by "a lift."

"I rather think I can," answered Neal; "but goodness! I feel as if there were aches and bruises all over me. Once or twice my head seemed jumping straight off my shoulders. No more going in a wagon over corduroy roads for me! I'd rather be leg-weary any day."

The travellers halted that evening about five o'clock on the banks of a lonely stream. The guides pitched the two tents — Joe had provided one for his party — facing each other on a patch of clearing, with a space of about fifteen feet between them, in the centre of which blazed a roaring camp-fire. Now all the axes and knifes among the band were in demand for cutting and sharpening stakes and ridge-poles on which to stretch their canvas.

Moreover, no evergreen boughs could be procured for beds; and the boys had to work with a will, helping Uncle Eb and Joe to cut bundles of the long, rank grass that grew by the water to form a bed for their tired bodies.

Every one was camp-hungry, as they had not halted for a meal since leaving the settlement. After a splendid supper of venison, broiled over sizzling logs, bread, and fried potatoes, — for they had added to their stores at the farm, — they had a glorious social hour by the camp-fire. Joe got off any amount of "ripping" stories; and the sound of many a jolly chorus, led by Cyrus, and swelled by the musical efforts of the entire crew, mingled with the lonely rustle of the night wind among faded and drifting leaves.

When Doc's summons came to turn in, they stretched themselves upon the grassy beds, not undressing, as the night was chilly and the temporary quarters were not so snug as their previous ones. Still in their warm jerseys, trousers, woollen stockings, and knitted caps, with the heat from the piled-up campfire streaming under the raised flaps of the tents, they slept as cosily as if they lay on spring mattresses, surrounded by pictured walls.

## CHAPTER XI.

#### BEAVER WORKS.

ABOUT noon on the following day they were obliged to bid farewell to Lin Hathaway, his wagon and horses, as the logging-road went no farther. The young settler turned homeward rather regretfully. It might be many months again before he got a chance of talking to anybody beyond his father and mother, and the boys had brought a dash of outside life into his woodland solitude.

The travellers proceeded on foot through a dense forest, which, luckily for Dol, had little undergrowth and mostly a soft carpet of moss or dry pine needles. Still they had plenty of climbing over windfalls, with many rough pokes and jibes from forward boughs and rotten limbs, to rob the way of sameness. Through this labyrinth they were safely piloted by Uncle Eb and Joe, the latter with his compass in his hand, and the former simply studying the "Indian's compass," which is observing how the moss grows upon the tree-trunks, there being always a greater quantity on the side which faces north.

Before nightfall they reached another log cabin, tenanted by a man who had just settled down for the purpose of clearing up a farm. Here they were lodged for the night, without trouble of making camp.

The third day of their journey was marked by two sensations. They halted for a short rest at a point where there was an extensive break in the forest. Scarcely had they emerged from the gloom of a dense growth of cedars, when Dol exclaimed, —

"Good gracious! That looks as if people had been building a jolly high railroad out here."

On the right rose a bare, steep ridge of sand and gravel, nearly ninety feet in height, and closely resembling a railway embankment.

"Well, boy," laughed Dr. Phil, "if that's a railroad, Nature built it, and by a mighty curious process too. The sand, rocks, and gravel of which it is mostly formed must have been swept here by a great rush of waters that once prevailed over this land. We call the ridge a 'Horseback.' If you like, we'll climb to the top of it, after we've had our snack [lunch], and you can get a peep at the surrounding country."

So they did. The top was level, and wide enough for two carriages to drive abreast; and the view from it was one which could never be forgotten. Around them were millions of acres of forest land, beautiful with the contrasts of October; here dipping into a cedar valley, in the midst of which they saw the silver smile of a woodland lake, there rising into a hill crowned with towering pines, some of them over a hundred feet in height.

But, most thrilling sight of all, they beheld, only half a dozen miles away, rising in sublime grandeur against the sky, the mountain of mountains in Maine, — great Katahdin. They had caught glimpses of its curved line of peaks before. Now they saw its forests, and the rugged slides where avalanches of bowlders and earth from the top had ploughed heavily downward, sweeping away all growth.

Cyrus lifted his hat, and waved it at the distant mass.

"Hurrah!" he cried. "There's the home of storms! There's old Katahdin! The Indians named it Ktaadn, 'the biggest mountain.'"

"Want to hear the Indian legend about it, lads?" asked Dr. Phil.

A general chirp of assent was his reply, and the doctor began:—

"Well, when the redskins owned these forests, they believed that the summit of Katahdin was the home of their evil spirit, or, as they call him, 'The Big Devil.' He was named Pamolah. And he was a mighty unpleasant sort of neighbor. Once, so tradition says, he ran away with a beautiful Indian maiden, and carried her up to his lonely lair among those peaks. When her tribe tried to rescue her, he let loose great storms upon them, his artillery being thunder, lightning, hail, and rain, before which they were forced to flee helter-skelter. An old red chief long ago told me the story, and added gravely that 'it was sartin true, for han'some squaw always catch 'em debil.'

"The foundation of the legend lies in the fact that there really is a very curious granite basin among Katahdin's peaks, and it is the birthplace of most storms which sweep over

IN THE SHADOW OF KATAHDIN.

our State. I myself have seen clouds forming in it, when I made an ascent of the mountain in my younger days, and whirling out in all directions. The roar of its winds may sometimes be heard miles away. There are several ponds in the basin; one of them, a tiny, clear lake, without any visible outlet, is Pamolah's fishing-ground. That's the yarn about the mountain as I heard it."

"Ain't it a'most time for us to be gittin' down from this Horseback, Doc?" asked Joe, who had been listening with the others. "I thought we'd reach the farm you're heading for to-night, but we're half a dozen miles off it yet; and we can't do more'n another mile or two afore it'll be time to halt and make camp. There's some pretty bad travelling and a plaguy bit of swamp ahead."

"I guess you're about right, Joe," said Doc, rising with alacrity from the stone where he had seated himself while telling his yarn.

Joe's bad travelling meant a great deal of tripping and floundering through soft mud and mire, with slippery moss-stones sandwiched in, and dwarfed bushes which ran along the ground, and twisted themselves in an almost impassable tangle. These had a knack of catching a fellow's feet, and causing him to sprawl forward on his face and hands, whereupon his knapsack would hit him an astounding thwack on the back.

After three-quarters of an hour of this fun, very muddy, clammy with perspiration, and thoroughly winded, the party reached firmer ground, and the guides called a halt.

"Guess we'd better rest a bit," said Joe, "afore we go farther. There's nothing in forest travelling that'll take the breath out of a man like crossing a swamp," eying compassionately the city folk; for he himself was as "fit" as when he started. "Then we'd better follow that stream till we strike a good place for a camping-ground. What say, Doc?"

Dr. Phil, as captain, signified his assent. After a short breathing-spell he again gave the command, "Forward!" And his company pushed on into the woods, following the course of a dark stream which had gurgled through the swamp.

"There used to be an old beaver-dam somewheres about here," broke forth Joe presently, when they had made about a quarter of a mile, the younger guide taking the lead, for he was evidently more at home in this part of the forest land than his senior, Uncle Eb. "Hullo, now! there it is. Look, gentlemen!"

He pointed to a curved bank of brush-wood, mostly alder branches, piled together in curious topsyturvy fashion, which formed a dam across the stream. It bristled with sticks, poking out and up in every direction; for the bushy ends of the boughs had been heavily plastered with mud and stones, to keep them down.

"That a beaver-dam!" gasped Neal in amazement. "Why, I always had an idea that beavers were half human in intelligence, and wove their branches in and out in a sort of neat basketwork when making dams. That's a funny rough-and-tumble looking old pile."

"It's a good water-tight dam, for all that," answered Cyrus. "And don't you begin to underrate Mr. Beaver's intelligence until you see more of his works. I've torn the bottom out of a dam like this on a cold, rainy night, — beavers like rainy nights for work, — and then hidden myself in some bushes to watch the result. It was a trial of strength and patience, I assure you, to remain there for six mortal hours, — though I had rubber over-alls on, — with wet twigs and leaves slapping my face. But the sight I saw was more wonderful than anything I could have imagined.

There was a cloudy, watery moon; and shortly after it rose, five beavers appeared upon the dam, scrambling up and down, and examining the great hole through which the water was fast leaking out of their pond. Then, following a big fellow, who was evidently the boss beaver, they swam to the bank. He stationed himself near a tree about twenty inches in circumference, and his four boys at once started to fell it. I tell you they worked like hustlers, each one sawing on it in turn with his sharp teeth, and sometimes two of them together on different parts of the trunk.

"At last the tree — it was an ash — fell, toppling into the water just where the beavers wanted it. They pushed and tugged it downstream for about ten yards, to the dam, and propped it against the opening which I had made. I couldn't see the rest of the operations clearly; but I caught glimpses of them, marching about on their hind-legs, carrying mud snug up to their chins like this," here Cyrus folded his arms across his chest. "And before daybreak that dam was perfectly repaired, with never a leak in it.

"You know they build the dams in very shallow water, only a few inches deep; and they generally roll in a couple of long logs for a solid foundation. It was one of these which I had torn out. Now, Neal, what do you say about the beaver's intelligence?"

"If I didn't know you, Cyrus, I'd say you were making up as you went along," answered Neal. " It seems one of those things which a fellow can scarcely believe in. Hulloa! What's that?"

A loud report, like the bang of a gun, made all the boys, who had been standing very quietly, gazing at the dam, suddenly jump.

"It's only a beaver striking the water with his tail," laughed Cyrus. "He has been swimming about somewhere up-stream, and has scented us, and dived. I have heard one do that a dozen times in the night, if he detected the presence of man; but it's very unusual in the daytime, for they rarely venture out in broad light. In diving, if suddenly alarmed, they strike the surface of the water a tremendous whack with their tails, as a signal of alarm, making this report, which in still weather resounds for a great distance.

"I'm very glad you heard it, boys; for your chances of seeing the master beaver or any of his colony are mighty slim. But we'll probably come on their lodge a little higher up."

Above the shallow water where the dam was built, the stream widened into a broad, deep pool. About fifty yards ahead, in the centre of this, was a tiny island. On its extreme edge Joe pointed out the beaver lodge. It was shaped something like a huge beehive, being about a dozen feet in diameter and five feet high. The outside seemed to be entirely covered with mud and fibrous roots, through which the sticks which formed its framework poked out here and there.

"The doors are all under water," said Cyrus, "and so far down that they'll be beneath the ice when the stream freezes in winter. Otherwise the beavers could not reach their pile of food-wood, which they keep at the bottom, and would starve to death. They are clerks of the weather, if you like. They seem to know when the first hard frost is coming, and sink their stores a day or two before. Man has not yet discovered their mysterious knack of sinking wood, and keeping it stationary through many months.

"They feed on the inner bark of poplar, white birch, and willow trees. In autumn they fell these along the banks, generally so that they will fall into the water, tug and push them down-stream, and float them near to their lodges. If the trees are too big to be easily handled, they saw them into convenient lengths."

"I call it tough luck, not being able to get a sight of the animals, after seeing so much of their works," grumbled Royal.

"Ye might wait here till midnight, and not have any better," said Joe. "That fellow's tail was like a fire-alarm to them. They ain't to home now, you bet! They've dusted out of their house as if it was on fire; and they've either dived to the bottom, or hidden themselves in holes along the bank. Guess we'd better be moving on. It's a'most time to think about making camp."

"The beavers have been working here!" exclaimed the guide a few minutes later, as he strode ahead. "These white birches were felled by 'em; and a dandy job they did too."

He pointed to two slim birches which lay prone with their tops in the water, and to a third, the trunk of which was partly sawn through in more than one place. The ground was strewn with little clippings of timber, bearing the saw-marks of the beavers' teeth. The boys gathered them up as curiosities.

"Oh, the skilful little animals can beat this work by long odds!" exclaimed Doc. "These trunks only measure from eight to twelve inches in circumference. I've seen a tree fully two feet round which was felled by them. Say, Joe! don't you think we'd better camp to-night somewhere on the *brûlée?*"

"Just what I'm planning, Doc," answered Joe. "We must be pretty near it now."

A few minutes afterwards the party filed out of the dense woods, passed through a grove of young spruces, forded a brook which emptied itself into the stream they were following, and came upon a scene blasted, barren, and unutterably dreary.

The band of boys, who, in spite of swamps and jungles, had learned to love the forest dearly, for its many beauties, and for the wild offspring with which it teemed, sorrowfully gasped, as if they saw the skeleton of a friend.

## CHAPTER XII.

### "GO IT, OLD BRUIN!"

BEFORE them lay a ruined tract of country, extending northward farther than eye could reach. It is called by Maine woodsmen a *brûlée*, a name borrowed from their French-Canadian neighbors, who dwell across the boundary line which separates the Dominion from the United States.

The word signifies "burnt tract;" but it gives a feeble idea of the fire-smitten, blackened region on which the lads looked.

The forest until now had been a wilderness truly, but a wilderness where every kind and size of growth, from the giant pine to the creeping wintergreen and shaded mosses, mingled in beautiful confusion. Here it became a desert. For the terrible forest fires, the woodsman's tragic enemy, had swept over it not long before, devastating an area of many square miles. Millions of dollars worth of valuable timber had been reduced to rotting embers. Storm-defying pines had crashed to the earth, and were overridden by the flames in their wild rush onward. Sometimes only a smutty stump showed where they had stood; sometimes, robbed of life and every limb, portions of the fire-eaten trunks still remained erect, — bare, blackened poles. All smaller growth, and even the surface of the ground, parched by summer heats, had burned like tinder. Rocks and stones were baked and crumbling.

"Boys, that's the most mournful sight a woodsman can see," said Doc, looking away over the wrecked region, touched with golden lights from an October sunset. "It makes one who loves the woods feel as if he had lost a living friend."

"Well, 'tain't no manner o' use to fret over it," declared Joe energetically. "Nature don't waste time in fretting, you bet! She starts in and tries to cover the stripped ground, as if she was sort of ashamed to have it seen."

The guide pointed earthward. At his feet

a dwarfed growth of blueberry bushes and tiny trees was already springing up to screen the unsightly, ash-strewn land.

"True enough, Joe! Nature is a grand one for remedies," answered the doctor. "Still, it will be half a century or more before she can raise a timber growth here again. Hulloa! Dol, what are you fellows up to?"

While his elders were studying the *brûlée*, Dol, who objected to dreary sights, had marched down to the brink of the stream, accompanied by Royal's young brothers, Will and Martin Sinclair. The little river gurgled and frisked along beside the burnt tract, like a line of life bordering death. It seemed to the boys to prattle about its victory over the flames when it stopped their sweeping course, so that the woods on its opposite bank were uninjured, as were those beyond the brook in the rear.

"We're studying the ways of the great sea-serpent!" shouted back Dol, who was splashing about in a sedgy pool.

By and by when the guides had finished their work of making camp, when they had pitched the tents, cut boughs for beds and fuel in the spruce grove behind, and were cooking an odorous supper, the three juveniles came slowly towards the camp-fire from the water.

"What on earth have you got there, young one?" asked Dr. Phil; for Adolphus Farrar was bareheaded, and carried his hat very gingerly, with its corners clutched together to form a bag.

"The big sea-serpent himself," answered Dol mysteriously.

Of a sudden he opened his dripping hat, and spilled out a small water-snake, about ten inches long, upon the doctor's lap.

There was a great roar of laughter, in which Dol's abettors, Will and Martin, joined with cheerful shouts. The little joke had the effect of winning everybody's thoughts from roaring flames, wrecked forests, and the dreary *brûlée*. Uncle Eb killed the snake, maintaining that water-snakes were "plaguy p'isonous," while Cyrus scouted the idea. The supper that evening was a merry enough meal. The camp, lit by the ruddy glow from its great fire, looked an oasis of light, warmth, and jollity in the black and burnt desert.

The darky, hearing Cyrus declare that he was fearfully hungry, mixed some flapjacks to form a second course, after the venison steaks and potatoes. He had exhausted his stock of maple sugar, but he produced a small wooden keg of the apparently inexhaustible molasses.

"He! he! he! Dat jest touches de spot, don't it?" he chuckled, when, having carefully served each member of the party, he seated himself about three feet from the camp-fire, with a round dozen of the thin cakes for his own eating.

He coated them with the thick molasses, and set the keg down side by side with a bag of potatoes which had been brought from the settlement.

There these provisions remained when, earlier than usual, the party turned in, and stretched their tired limbs to rest, lying down, as they had done before when sleeping under canvas, with all their garments on save coats and moccasins. Whether Uncle Eb forgot his "m'lasses," or whether he purposely left it without, there not being a spare inch of room in the small tents, no one then or afterwards inquired.

As a result of the jolly intimacy that had sprung up between the two companies during the few days when they had all things in common, the boys disposed of themselves for the night as they pleased. Neal turned in with the doctor, Royal, and Joe, the four stretching themselves on the evergreen boughs, with their feet to the opening of the tent, and their rifles and ammunition within reach. Of course the Winchesters were empty, it being a strict rule that firearms should not be brought into camp loaded.

The younger Sinclairs, with Cyrus, Dol, and Uncle Eb, occupied the other tent.

It seemed to Neal that he had hardly slept one hour, — probably it was nearer to three, — during which time he had been dreaming with vague foreshadowings of the final and crowning sport of the trip, the grand moose-stalking, and of Herb Heal, the mighty hunter, when he was awakened by a shrill scream just outside the canvas. He started, with his heart going whackety-whack. The cry was sudden and intensely startling, appearing twice as loud as it really was when it broke the pathetic stillness of the *brûlée*, where not a tree rustled or twig snapped, and the night wind only sighed faintly and fitfully through the newly springing growth.

Again sounded that startling screech; and yet again, making a dreary, piercing din.

"By all that's funny! it's another coon,"

gasped Neal; and he gently pinched the shoulder of Joe, who lay on his left.

"Joe!" he whispered. "Wake up! There's a raccoon just outside the tent. I heard his cry."

The guide was awake and alert in an instant. So, too, was Dr. Phil.

"What's up, boys?" asked the latter, hearing a murmur.

"There's a coon close by," said Neal again. "Listen to him!"

Even while he spoke, young Farrar caught sight of two feathered things hopping along the avenue of light which lay between him and the camp-fire, the red flare of the flames mingling with the white radiance of a cloudless moon. At the same time the screech sounded and resounded.

"Coon!" exclaimed Joe derisively. "That's no coon. It's only a little owl. Bless ye! I've had five or six of 'em come right into this tent of a night, and ding away at me till I had to talk to 'em with the rifle to scare 'em off. I'll give 'em a dose o' lead now if they don't scoot mighty quick; that'll stop their song an' dance."

"Their cry is pretty much like a raccoon's, Neal," said Doc. "Only it's a great deal weaker. Lie down, boy. Go to sleep, and don't mind them."

The owls perhaps apprehended danger. At all events, they were silent for a while; and in three minutes each occupant of the tent was fast asleep again, with the exception of Neal. The sharp awakening had upset his nerves a bit. He obeyed the doctor, and hugged his blankets round him, hoping sleep would return; but he lay with eyes narrowed into two slits, peeping at the ruddy camp-fire, involuntarily listening for the screeching of the birds, and wishing that he had not been such a greenhorn as to disturb his comrades for nothing. Royal, who lay on his right, was of a less excitable temperament. Although he had been awakened, he was now snoring lustily, insomnia being a rare affliction in camps.

"What's that?"

About half an hour had passed when Neal Farrar suddenly and sharply rapped out these words close to Joe's ear. He felt certain that he would not now bring upon him the woodsman's good-natured scorn for making a disturbance about nothing. A heavy, stealthy tread, as of some big animal, was crushing the pygmy bushes near the tent. Immediately afterwards he saw an uncouth black shape in the lane of light between himself and the fire. It disappeared while his heart was giving one jump, and he heard a dull, mumbling noise, such as a pig might make when rooting amid rubbish, varied with an occasional low growl.

Joe was already awake. His hunter's instinct told him that something truly exciting was on now.

"My cracky! I b'lieve it's a bear!" he muttered, forming his words away down in his throat, so that Neal only caught the last one. "Keep still as death!"

The guide reached out a long arm, and clutched his rifle. Hurriedly he jammed half a dozen cartridges into its magazine. Then lightly and silently, as if he was made of cork, he got upon his feet, and bounded out of the tent, Neal copying his actions nimbly and noiselessly as he could; though, in his excitement, he only succeeded in getting two cartridges into his Winchester.

Royal's snoring ceased. Doc's eager question, "What's up now, boys?" reached the two just as they quitted shelter, and passed into the broad moonlight, crossed with red gleams from their fire.

"A bear!" yelled Joe in answer, his rifle and he breaking silence together.

Three times the Winchester sharply cracked.

Then with a mad "Halloo!" the guide seized a flaming stick from the fire, and, swinging it above his head, started after the big black animal of which Neal had caught a glimpse before. He now saw it plainly as, already fifty yards ahead, it made off at a plunging gallop across the moonlit *brûlée*.

Young Farrar had been the champion runner of his school, and he blessed his trained legs for giving him a prominent part in the wild chase that followed. Still imitating the woodsman, he pulled another half-lighted stick from the camp-fire, and waved it in a frenzy of excitement, while he ran like a buck at Joe's side.

"Tumble out! Tumble out, boys! A bear! A bear!" now rang from one tent to another.

In two minutes every camper, in his stocking feet, just as he had risen from his bed, was tearing across the *brûlée* in the wake of Bruin, yelling, leaping, and swinging smouldering firebrands.

It was a scene and a chase such as the boys, in their most far-fetched dreams, had never pictured, — the white moonlight glimmering on the black stumps and tottering trunks of the ruined tract, the hunted bear plunging off among them, frightened by the shouting and the lights, the heavy, lumbering gallop enabling it at first to distance its pursuers.

Owing to their fleetness and the odds they had at the start, the guide and Neal kept far ahead of their comrades. The noise which Bruin made as he lumbered over the pygmy growth, and the charred, rotting timber that littered the ground beneath it, were quiet enough to guide Joe unerringly in the bear's wake, even when that bulky shape was not distinguishable.

"What's this?" screeched the woodsman suddenly, as he stumbled upon something at his feet. "By gracious! it's our keg of m'lasses. He made off with that, and has dropped it out o' sheer fright, or because he's weakening. I know I hit him twice when I fired; but he's not hurt too badly to run, or to fight like a fiend if we come to close quarters. Like as not 'twill be a narrow squeak with us if we tackle him. If you're scared a little bit, Neal, let up, an' I'll finish him alone."

"Scared!" Neal flung the word back with scorn, as if he was returning a blow.

For the life of him he could not bring out another syllable, going at a faster rate than ever he had done in the most stubbornly contested handicap. The strong-winded guide rapped out his sentences as he ran, apparently without waste of breath.

The feverish enthusiasm of the hunter, which he had never felt before, was now alive in Neal. His blood raced through his veins like liquid fire. He had been long enough in Maine to know that in wreaking vengeance on Bruin for many misdeeds he would be acting in the interests of justice. For the black bear is still such a master pest to the settlers who are trying to establish their farms amid the forests where it roams, that the State has outlawed the beast, and pays a bounty for its skin.

Joe thought little about this; for a gentleman whom he had guided early in the summer had lately written to him, offering a price of fifteen dollars for a good bearskin.

Here was the woodsman's golden opportunity — an opportunity for which he had been thirsting since the receipt of that letter.

He already regarded his triumph over the bear as secure, and its hide as forfeited. He nearly caused Neal Farrar to burst a blood-

"Go it, Old Bruin! Go it while You can!"

vessel from the combined effects of struggling laughter and running, when he began to apostrophize the flying foe with grim humor, thus:—

"Go it, old Bruin! Go it while ye can! There ain't a hair on yer back that b'longs to ye!"

But it soon became evident that the bear couldn't go on much longer at this breakneck pace. Its pursuers heard its steps with increasing distinctness, and then its labored breathing. They were gaining on it fast.

The brute came into full view about forty yards ahead, as it ascended a slight elevation, crowned with blasted tree trunks.

"I'll draw bead on him from here," said Joe, stopping short. "Get ready to fire, lad, if he turns. It'll take lots o' lead to finish that fellow."

Twice Joe's rifle spoke again. One shot took effect. There was a fearful growl from the beast, but it was not yet mortally wounded.

Maddened and desperate, it wheeled about, and came straight for its pursuers. Again the guide fired. Still the bear advanced, gnashing its teeth and mumbling horribly; Neal saw its black shape not thirty yards from him.

"Shoot! shoot, boy!" screamed Joe. "Or give me your rifle. I haven't got a charge left!"

For half a minute Farrar shook all over as with ague. His nostrils felt choked. His mouth was wide open in his efforts to breathe. His heart pounded like a sledge-hammer. With that mumbling brute advancing upon him, he felt as if he couldn't fire so as to hit a haystack or a flock of hens at a barn-door.

Then, suddenly, he was cool again, seeing and hearing with extraordinary clearness. The ignominious alternative of giving his rifle to Joe produced a revulsion. His fingers were on the trigger, his left hand firmly gripped the barrel of his Winchester; he brought it to his shoulder.

"Aim low! Try to hit him in the front of the neck where it joins the body," said Joe, in tones sharp as a razor, which cut his meaning into Neal's brain.

Bruin was only fifteen yards away when Farrar's rifle cracked once — twice — sending out its messengers of death.

There was a last terrible growl, a plunge, and a thud which seemed to shake the ground under Neal's feet. As the smoke of his shots cleared away, Joe beheld him leaning on his rifle, with a face which in the moonlight looked white as chalk, and the bear lying where it had fallen headlong towards him. It made a desperate struggle to regain its feet, then rolled on its side, dead.

One bullet had pierced the spot which Joe mentioned, and had passed through the region of the heart.

## CHAPTER XIII.

### "THE SKIN IS YOURS."

A REGULAR war-dance was performed about the slain marauder by the young Sinclairs and Dol Farrar, when these laggards in the chase reached the spot where he fell. The firebrands had all died out before the enemy turned; but in the white moon-radiance the bear was seen to be a big one, with an uncommonly fine skin.

Neal took no part in the triumphal capers. He still leaned upon his rifle, his breath coming in gusty puffs through his nostrils and mouth. Not alone the desperate sensations of those moments when he had faced the gnashing, mumbling brute, but the unexpected success of his first shot at big game, had unhinged him. By his endurance in the chase, by the pluck with which he stood up to the bear, above all, by his being able, as Joe phrased it, to "take a sure pull on the beast at a paralyzing moment," he had eternally justified his right to the title of sportsman in the eyes of the natives. The guides, Joe and Eb, were not slow in telling him that he had behaved from start to finish like no "greenhorn," but a regular "old sport."

"My cracky! 'twas lucky for me that you had game blood in you, which showed up," exclaimed Joe, catching the boy's arm in a friendly grip, with an odd respect in his touch, which marked the admission of young Farrar into the brotherhood of hunters. "I hadn't a charge left, an' not even my hunting-knife. Lots o' city swells 'u'd have been plumb scared before a growler like that," — touching Bruin's carcass with his foot, — "even if they had a small arsenal to back 'em up. They'd have dropped rifle and cartridges, and hugged the nearest trunk. I've seen fellers do it scores o' times, bless ye! after they came out here rigged up in sporting-book style, talking fire about hunting bears and moose. But that was all the fire there was to 'em."

Yet Neal's triumph over the poor brute, which had raced well for its life, was not without a faint twinge of pain; and he was too manly to look on this as a weakness. A sportsman he might be, of the sort who can shoot straight when necessity demands it, but never of that class who prowl through the forests with fingers tingling to pull the trigger, dreading to lose a chance of "letting blood" from any slim-legged moose or velvet-nosed buck which may run their way. It needed Doc's praise to make him feel fully satisfied with his deed.

"It was a crack shot, boy," said the doctor proudly. "And I guess the farmer at the next settlement will feel like giving you a medal for it. Old Bruin has only got what he gave to every creature he could master."

There being no tree conveniently near to which they could string up the dead bear, the guides decided to leave the ugly matter of skinning and dissecting him for morning light. The excited party returned to camp, but not to sleep. They built up their scattered fire, squatted round it, and discoursed of the night's adventure until a clear dawn-gleam brightened the eastern sky. Then Uncle Eb and Joe started out again across the *brûlée*. They reappeared before breakfast-time, bringing Bruin's skin and a goodly portion of his meat.

Joe laid the hide at Neal's feet.

"There, boy," he said, "the skin is yours. It belongs rightly to the man who killed the bear; and I guess the brute wasn't mortally hurt at all till your bullet nipped him in the neck."

"But what about the fifteen dollars from that New York man, Joe? You'll lose it," faltered young Farrar, with a triumphant heart-leap at the thought of taking this trophy back to England, but loath to profit by the woodsman's generosity.

"Don't you bother about that; let it go," answered Joe, whose business of guiding was profitable enough for him. "'Tain't enough for the skin, anyhow. Nary a finer one has been taken out o' Maine in the last five years; and mighty lucky you Britishers were to git a chance of a bear-hunt at all. Old Bruin must have been powerful hungry to come around our camp."

There was a grand breakfast before the travellers broke camp that morning. The guides and Doc — who had got accustomed to the luxury during visits to settlers and lumber-camps — feasted off bear-steaks. Cyrus and the boys, American and English, declined to touch it. The whole appearance of Bruin as he lay stretched on the ground the night before made their "department of the interior" revolt against it.

When a start was made for the settlement, Joe bundled up the skin, and, as a tribute of respect to Neal's "game blood," carried it, in addition to his heavy pack, for a distance of four miles over the desolate *brûlée* and across a soft, miry bog. On reaching the farm clearing, he cut the stem of a tall cedar bush, which he bent into the shape of a hoop, binding the ends together with cedar bark. He then pricked holes all around the edges of the hide with the sharp point of his hunting-knife, stretched it to its full extent, and fastened it to the hoop, which he hung up to a tree near the settler's cabin, telling Neal that in a few days it would be dry enough to pack away in a bag.

But as it was a cumbersome article to carry while tramping a dozen miles farther to the camp on Millinokett Lake, the farmer offered to take charge of it for its owner until he passed that way again on his return journey; an offer which Neal thankfully accepted. The old backwoodsman was, truth to tell, delighted to see hanging up near his cabin door the skin of an enemy who had ofttimes plundered him so unmercifully.

He made the travellers royally welcome, let them have the roomy kitchen of his log shanty to sleep in, with a soft bed of hay. Here he lay with them, while his wife and sickly little girl occupied an adjoining space about twelve feet square, which had been boarded off. This was all the accommodation the log home afforded.

The forest child was a puzzle to the lads. To them she looked as if the soul of a grandmother had taken possession of a thin, long-limbed body which ought to belong to a girl of ten. Her pinched features and over-wise eyes told a tale of suffering, and so did her high-pitched, quivering voice, as it made elfishly sharp remarks about the boys until they blenched before her.

This was the little one of whom the doctor had said "that she fretted if he did not come to see her once in a while." And with Doc she was a different being. Her voice softened, her eyes became childlike, and thin tinkles of laughter broke from her as she clung to him, and received certain presents of medicines and picture-books which he had brought for her in a corner of his knapsack.

For two nights the travellers slept in a row on their hay bed; for two long-remembered days the five boys roamed the country round the clearing, starting deer, catching glimpses of a wildcat, a marten or two, and of another coon. Then came, to use Dol's expression, "the beastly nuisance of saying good-by."

Dr. Phil was obliged to return to Greenville; and he declared that now he must surely start his nephews homeward, for Royal expected to graduate from the High School during the following year, and to let him waste more time from study would be questionable kindness. Joe Flint of course would go back with his party. And here Cyrus paid Uncle Eb's fees for guiding, and dismissed him too.

Only a dozen miles of tolerably easy travelling now separated Garst and his English comrades from the camp on Millinokett Lake, where they were to meet the redoubtable Herb Heal. The settler, knowing this tract of country as thoroughly as he knew his own few fields, offered to lead our trio for the first half of their onward march; and as they could follow a plain trail for the remainder of the way, they had no further need of their guide's services. They promised to visit Eb at his bark hut on their return journey, to bid him a final farewell, and hear one more stave of: —

"Ketch him, Tiger, ketch him!"

"Good-by, you lucky fellows!" said Royal Sinclair huskily, as he gripped Neal's hand, then Dol's, in a brotherly squeeze when the hour of parting came. "I wish I was going on with you. We've had a stunning good time together, haven't we? And we'll run across each other in these woods some time or other again, I know! You'll never feel satisfied to stay in England, where there's nothing to hunt but hares and foxes, after chasing bears and moose."

"Oh! we'll come out here again, depend upon it," answered Neal. "Drop me a line occasionally, won't you, Roy? Here's our Manchester address."

"I will, if you'll do the same."

"Agreed. Good-by again, old fellow!"

"I've got the slip of birch-bark and the horn safe in my knapsack, Doc," Dol was saying meanwhile, feeling his eyes getting leaky as he bade farewell to the doctor. "I — I'll keep them as long as I live."

Doctor Phil had been as good as his word. He had made Joe rip the slip of white bark, with the rude writing on it, off the pine-tree near the swamp, and had presented it to Dol ere the boy quitted his camp.

"Well, confusion to partings anyhow!" broke in Joe. "Don't like 'em a bit. Hope you'll get that bear-skin safe to England, Neal. When you show it to your folks at home, tell 'em Joe Flint said he knew one Britisher who would make a woodsman if he got a chance. Don't you forgit it."

"Good-by," said the doctor, as he clasped in turn the hands of the departing three. "Good luck to you, boys! Keep your souls as straight as your bodies, and you'll be a trio worth knowing. We'll meet again some day; I'm sure of it."

Martin and Will were chirping farewells, and lamenting that they would have no more chances of studying water-snakes in sedgy pools with Dol. Amid cheers and waving of hats the campers separated.

"Forward, Company Three!" cried Cyrus encouragingly, stepping briskly ahead, his comrades following. "Now for a sight of the 'Jabberwock' of the forest, the mighty moose. Hurrah for the wild woods and all woodsmen!"

## CHAPTER XIV.

### A LUCKY HUNTER.

AMID cracking of jokes, and noise which would have disgraced a squad of Indians, "Company Three," as Cyrus dubbed his reduced band, reached the crowning-point of their journey, the log camp on the shore of Millinokett Lake.

During the first half-dozen miles of the way, though each one manfully did his best to be lively, a sense of loss made their fun flat and pointless. Royal's tear-away tongue, his brothers' racket, Joe's racy talk, Uncle Eb's kind, dark face, and more than all, Doc's companionship, which was as tonic to the hearts of those who travelled with him, were missed.

But spirits must be elastic in forest air. When they halted at noon to eat their "snack" on the side of a breezy knoll, with a tiny brook purling through a pine grove beneath them, with Katahdin's rugged sides and cloud-veiled peaks looming in majesty to the north, the thought of what lay behind was inevitably lost in what lay before. Enthusiasm replaced depression.

"It's no use grizzling because we can't have those fellows with us all the time," remarked Neal philosophically. "'Twas a big piece of luck our running against them at all. And I've a sort of feeling that this won't be the end of it; we'll come across them again some day or other."

"And at all events we'll probably get a sight of Doc at Greenville as we go back," said Dol, to whom this was no small comfort.

"Well, needless to say, I'd have been glad of their company for the rest of the trip. But still, if they had taken a notion to come on with us, it would have reduced to nothing our chances of seeing a moose. We're a big party already for moose-calling or stalking — three of us, with Herb;" this from Cyrus.

"Now, fellows, don't you think we'd better

get a move on us?" added the leader. "We've half a dozen miles to do yet; but the trail begins right here, and is clearly blazed all the way to our camp. Let's keep a stiff upper lip, and the journey will soon be over."

It was very delightful to sit there in the crisp October air, with the brook seemingly humming tender legends of the woods, which witless men could not translate, with an uncertain breeze playing through the newly fallen maple-leaves, now turning them one by one in lazy curiosity, then of a sudden making them caper and swirl in a scarlet merry-go-round. Still, the young Farrars were not loath to move on. Now that they were nearing the climax of their journey, their minds were full of Herb Heal. Their longing to meet this lucky hunter grew with each mile which drew them nearer to him.

They pressed hard after their leader, looking neither right nor left, while he carefully followed the trail; and one hour's tramping brought them to the shores of Millinokett Lake.

Here, despite their eagerness to reach their new camp, they were forced to stop and admire the great sheet of forest-bound water, smiling back the sky in tints of turquoise and pearl, dotted with apparently countless islets, like specks upon the face of a mirror.

The irregular shores of the lake were broken by "logons," narrow little bays curving into the land, shining arms of water, sometimes bordered by evergreens, sometimes by graceful poplars and birches. From the opposite bank the woods stretched away in undulating waves of ridge and valley to the foot of Mount Katahdin, which still showed grandly to the northward.

"Millinokett Lake," said Cyrus, prolonging the syllables with a soft, liquid sound. "It's an Indian name, boys; it signifies 'Lake of Islands.' Whatever else the red men can boast of, the music of their names is unequalled. I don't know exactly how many of those islets there are, but I believe Millinokett has over two hundred of them anyhow. Our camp is on the western shore. Shall we be moving?"

After skirting the water for another mile or two, the travellers reached a broad, open tract, bare of timber. At the farther end of this clearing were two log cabins, low, but very roomy, situated at a distance of a few hundred yards from the lake, with a background of splendid firs and spruces, the lively green of the latter making the former look black in contrast.

"Is that our camp? How perfectly glorious!" boomed Neal and Dol together.

"It's our camp, sure enough," answered Garst, with no less enthusiasm. "At least the first cabin will be ours. I don't know whether there are any hunters in the other one just now."

The log shanties had been put up by an enterprising settler to accommodate sportsmen who might penetrate to this far part of the wilds in search of moose or caribou. Cyrus had arranged for the use of one during the months of October and November. Here it was that Herb Heal had engaged to await him. And as he had commissioned this famous guide to stock the camp with all such provisions as could be procured from neighboring settlements, such as flour, potatoes, pork, etc., he expected to slide into the lap of luxury.

In one sense he did. When the trio, their hearts thumping with anticipation, reached the low door of the first cabin, they found it securely fastened on the outside, so that no burglar-beast could force an entrance, but easily opened by man. Cyrus hurriedly undid the bolts, and stepped under the log roof, followed by his comrades. The camp was in beautiful order, clean, well-stocked, and provided with primitive comforts. An enticing-looking bed of fresh fir-boughs was arranged in a sort of rude bunk which extended along one side of the cabin, having a head-board and foot-board. The latter was fitted to form a bench as well. A man might perch on it, and stretch his toes to the fire in the great stone fireplace only two feet distant.

The boys could well imagine that this would make an ideal seat for a hunter at night, where he might lazily fill his pipe and tell big yarns, while the winter storm howled outside, and snow-flurries drifted against his log walls. But they looked at it wistfully now, for it was empty. There was no figure of a moccasined forest hero on bench or in bunk. There was no Herb Heal.

"Bless the fellow! Where on earth is he?" Garst exclaimed. "He's been here, you see, and has the camp provisioned and ready. Perhaps he's only prowling about in the woods near. I'll give him a 'Coo-hoo!'"

He stepped forth from the cabin to the middle of the clearing, and sent his voice

"Herb Heal."

ringing out in a distance-piercing hail. He loaded his rifle and blazed away with it, firing a volley of signal-shots.

Neither shout nor shots brought him any answer.

The second cabin was likewise empty, and, judging from the withered remains of a bed, had evidently been long unused.

"Well, fellows!" said the leader, with manifest chagrin, "we'll only have to fix up something to eat, make ourselves comfortable, and wait patiently until our guide puts in an appearance. Herb Heal never broke an engagement yet. He's as faithful a fellow as ever made camp or spotted a trail in these forests. And he promised to wait for me here from the first of October, as it was uncertain when I might arrive. I'm mighty hungry. Who'll go and fetch some water from the lake while I turn cook?"

Dol volunteered for this business, and brought a kettle from the cabin. He found it near the hearth, on which a fire still flickered, side by side with a frying-pan and various articles of tinware. Cyrus rolled up his sleeves, took the canisters of tea and coffee with other small stores from his knapsack, proceeded to mix a batter for flapjacks, and showed himself to be a genius with the pan.

The meal was soon ready. The food might be a little salt and greasy; but camp-hunger, after a tramp of a dozen miles, is not dulled by such trifles. The trio ate joyously, washing the fare down with big draughts of tea, rather fussily prepared by Neal, which might have "done credit to many a Boston woman's afternoon tea-table" — so young Garst said.

Yet from time to time longing looks were cast at the low camp-door. And when daylight waned, when stars began to glint in a sky which was a mixture of soft grays and downy whites like a dove's plumage, when the islets on Millinokett's bosom became black dots on a slate-gray sheet, and no laden hunter with rifle and game put in an appearance, even Cyrus became fidgety and anxious.

"I hope the fellow hasn't come to grief somewhere in the woods," he said, while a shiver of apprehension shot down his back. "But Herb has had so many hairbreadth escapes that I believe the animal has yet to be born which could get the better of him. And he can find his way anywhere without a compass. Every handful of moss on a trunk or stone, every turn of a woodland stream, every sun-ray which strikes him through the trees, every glimpse of the stars at night, has a meaning for him. He reads the forest like a book. No fear of his getting lost anyhow. Come, boys, I guess we'd better build up our fire, make things snug for the night, and turn in."

Rather dejectedly the trio set about these preparations. In twenty minutes' time they were stretched side by side in the wide bunk, with their blankets cuddled round them, already venting random snores.

"Hello! So you've got here at last, have you?"

The exclamations were loud and snappy, and awoke the sleeping campers like the banging of rifle-shots. With jumping pulses they sprang up, feeling a wave of cold air sweep their faces; for the cabin-door, which they had closed ere lying down, was now ajar.

The camp was almost in darkness. Only one dull, red ray stole out from the fire, on which fresh logs had been piled. But while the young Farrars rubbed their sleep-dimmed eyes, and slowly realized that the woodsman whom they had been expecting had at last arrived, a strangely brilliant illumination lit up the log walls.

This sudden and bewildering light showed them the figure of a hunter in mud-spattered gray trousers, with coarse woollen stockings of lighter hue drawn over them above his buckskin moccasins. His battered felt hat was pushed back from his forehead, a guide's leathern wallet was slung round him, and the rough, clinging jersey he wore, being stretched so tightly over his swelling muscles that its yarn could not hold together, had a rent on one shoulder.

His slate-gray eyes with jetty pupils, which were miniatures of Millinokett Lake at this hour, gazed at the awakened trio in the bunk, with a gleam of light shooting athwart them, like a moonbeam crossing the face of the lake.

The hunter held in his hand a big roll of the inflammable paper-like bark of the white birch-tree, which he had brought in with him to kindle his fire, expecting that it had gone out during his absence. Seeing a glow still on the hearth, and feeling instantly that the cabin was tenanted, he had applied a match to his bark, causing the vivid flare which revealed him to the eyes of those who had longed for his presence.

"Herb Heal, man, is it you?" shouted Cyrus, his voice like a midnight joy-chime, as he sprang from the fir-boughs and gripped the woodsman's arm. "I'm delighted to see you, though I was ready to swear you wouldn't disappoint us! I didn't fasten the cabin-door, for I thought you might possibly get back to camp during the night."

"Cyrus, old fellow, how goes it?" was Herb's greeting. "I had a'most given up looking for you. But I'm powerful glad you've got here at last."

The hunter's voice had still the quick snap and force which made it startling as a rifle-shot when he entered the cabin.

"These are my friends, Neal and Adolphus Farrar," said Cyrus, introducing the blanketed youths, who had now risen to their feet. "Boys, this is Herb Heal, our new guide, christened Herbert Healy — isn't that so, Herb?"

"I reckon it is;" answered the young hunter, laughing. "But no woodsman could spring a sugary, city-sounding name like that on me. I've been Herb Heal from the day I could handle a rifle."

He nodded pleasantly as he spoke to the strange lads, and began to chat with them in prompt familiarity, looking straight and strong as a young pine-tree in the halo of his birch torch. Garst, whose inches his juniors had hitherto coveted, was but a stripling beside Herb Heal.

"Is this your first trip into Maine woods, younkers?" he asked. "Well, I guess you've come to the right place for sport. I'm sorry I wasn't on hand to welcome you when you arrived. A pretty forest guide you must have thought me. But I guess I'll show you a sight to-morrow that'll wipe out all scores."

There was such triumph in the hunter's eye that the voices of the trio blended into one as they breathlessly asked, —

"What sight is it?"

"A dead king o' the woods, boys," answered Herb Heal, his voice vibrating. "A fine young bull-moose, as sure as this is a land of liberty. I dropped him by a logon on the east bank of Fir Pond, about four miles from here. I started out early, hoping to nab a deer; for I had no fresh meat left, and I didn't want to have a bare larder when you fellows came along. But the woods were awful still. There didn't seem to be anything bigger than a field-mouse travelling. Then all of a sudden I heard a tormented grunting, and the moose came tearing right onto me. I was to leeward of him, so he couldn't get my scent. A man's gun doesn't take long to fly into position at such times, and I dropped him with two shots. There he lies now by the water, for I couldn't get him back to camp till morning. He's not full-grown; but he's a fine fellow for all that, and has a dandy pair of antlers. By George! I'd give the biggest guide's fees I ever got if you fellows had been there to hear him striking the trees with 'em as he tore along. He was a buster.

"But you'll see him to-morrow anyhow, and have a taste of moose-meat for the first time in your lives, I guess."

Here Herb waved the fag-end of his bark roll, threw it down as it scorched his horny fingers, and stamped upon it.

The interior of the log cabin, ere it was extinguished, was a scene for a painter, — the lithe, muscular figure, tanned face, and gleaming eyes of the lucky hunter shown by the flare of his birch torch, and the three staring listeners, with blankets draped about them, who feared to miss one point of his story.

Cyrus was grinding his teeth in vexation

that he had narrowly missed seeing the moose alive. The two Farrars were burning with excitement at the thought of beholding the monarch of the forest at all, even in death. For they had heard enough wood-lore to know that the bull-moose, with his extreme caution, is like a tantalizing phantom to hunters. Continually he lures them to disappointment by his uncouth noises, or by a sight of his freshly made tracks, while his sensitive ears and super-sensitive nose, which can discriminate between the smell of man and every other smell on earth, will generally lead him off like a wind-gust before man gets a sight of him.

"I'm sorry to keep you awake, boys," said Herb Heal, making for the fire, after he had finished his story; "but I haven't had a bite since morning, and I'm that hungry I could chaw my moccasins. I'll get something to eat, and then we'll turn in. We'll have mighty hard work to-morrow, getting the moose to camp."

Herb was not long in making ready the stereotyped camp-fare of flapjacks and pork. To light his preparations, he took a candle out of a precious bundle which he had brought from a town a hundred miles distant, and set it in a primitive candlestick. This was simply a long stick of white spruce wood, one end of which was pointed, and stuck into the ground; the other was split, and into it the candle was inserted, the elasticity of the fresh wood keeping the light in place.

The tired hunter did not dawdle over his supper. In a quarter of an hour he had finished it, and was building up the fire again. Then he stretched himself beside the trio in the rude bunk, drawing one thin blanket over him. Neal, who lay on his right, was conscious of some prickings of excitement at having such a bedfellow on the fir-boughs, — the camper's couch which levels all. There flashed upon the fair-haired English boy a remembrance of how Cyrus had once said that "in the woods manhood is the only passport." He thought that, measured by this standard, Herb Heal had truly a royal charter, and might be a president of the forest land; for he looked as free, strong, and unconquerable as the forest wind.

## CHAPTER XV.

#### A FALLEN KING.

THE hunter was the only one who slept soundly that night on the fragrant boughs. Nevertheless, the moose was on his mind. Again in his dreams he imagined himself back by the quiet, shining logon, listening to the ring of the antlers as they struck the trees, and to the heaving snorts and deep grunts of the noble game as it tore through the forest to its death.

The moose was on the minds of his companions too. Again and again they awoke, and pictured him lying by the pond, where he had fallen, — a dead monarch. They tossed and grumbled, longing for day.

Neal and Dol surprised themselves and

their elders by being up and dressed shortly after five, before a streak of light had entered the cabin. But their guide was not much behind them. Herb had the camp-fire going well, and was preparing breakfast before six o'clock. The campers tucked away a substantial meal of fried pork, potatoes, and coffee. The first glories of the young sun fell on their way as they started across the clearing and away through the woods beyond, towards the distant pond where the hunter had got his moose.

Lying amid the small growth and grasses, by a lonely, glinting logon, they found the conquered king, sleeping that sleep from which never sun again would wake him. A bullet-hole, crusted with dark blood, showed in his side. The slim legs were bent and stiff, and the mighty forefeet could no more strike a ripping blow which would end a man's hunting forever. The antlers which had made the forest ring were powerless horn.

"Do you know, boys," said Herb, as he stooped and touched them, fingering each prong, "I've hunted moose in fall and winter since I was first introduced to a rifle. I've still-hunted 'em, called 'em, and followed 'em on snowshoes; but I never felt so thundering mean about killing an animal as I did about dropping this fellow. After his antics in the woods, when he tramped out onto the open patch where I was waiting under cover of those shrubs, I popped up and covered him with my Winchester. He just raised the hair on his back and looked at me, with a way wild animals sometimes have, as if I was a bad riddle. Like as not he'd never seen a human being before, and a moose's eyes ain't good for much as danger-signals. It's only when he hears or smells mischief that he gets mad scared.

"Well, I was out for meat, and bound to have it; so I pulled the trigger, and killed him with two shots. When the first bullet stung him he reared up, making a sharp noise like a wounded horse. Then he swung round as if to bolt; but the second went straight through his heart, and he fell where you see him now. I made sure that he was past kicking, and crept close to his head, thinking he was dead. He wasn't quite gone, though; for he saw me, and laid back his ears, the last pitiful sign a moose makes when a hunter gets the better of him. I tell you it made me feel bad — just for a minute. I've got my

A FALLEN KING.

moose for this season, and I'm sort o' glad that the law won't let me kill another unless it's a life-saving matter."

"How tall should you say this fellow was when alive?" asked Cyrus, stroking the creature's shaggy hair, which was a rusty black in color.

"Oh! I guess he stood about as high as a good-sized pony. But I've shot moose which were taller than any horse. The biggest one I ever killed measured between seven and eight feet from the points of his hoofs to his shoulders, and the antlers were four feet and nine inches from tip to tip. He was a monster — a regular jing-swizzler! A mighty queer way I got him too! I'll tell you all about it some other time."

"Oh! you must," answered Garst. "You'll have to give us no end of moose-talk by the camp-fire of evenings. These English fellows want to learn all they can about the finest game on our continent before they go home."

"Why, for evermore!" gasped Herb, in broad amazement. "Are you Britishers? And have you crossed the ocean to chase moose in Maine woods? My word! You're a gamy pair of kids. We'll have to try to accommodate you with a sight of a moose at any rate — a live one."

Though they would gladly have appropriated the compliment, the "gamy kids" were obliged to acknowledge that hunting had not been in their thoughts when they traversed the Atlantic. But they avowed that they were the luckiest fellows alive, and that the American forest-land, with its camps and trails and wild offspring, was such a glorious old playground that they would never stop singing its praises until a swarm of boys from English soil had tasted the novel pleasures which they enjoyed.

"Now, then, gentlemen!" said the guide, "I haven't much idea that we'll be able to haul this moose along to camp whole. If I skin and dress him here, are you all ready to help in carrying home the meat?"

The trio briskly expressed their willingness, and Herb began the dissecting business; while from a tree near by that strange bird which hunters call the "moose-bird" screamed its shrill "What cheer? What cheer?" with ceaseless persistence.

"Oh, hold your noise, you squalling thing!" said the guide, answering it back. "It's good cheer this time. We'll have a feast of moose-meat to-night, and there'll be pickings for you."

He then explained, for the benefit of the English lads, that this bird, whose cry is startlingly like the hunters' translation of it, haunts the spot where a moose has been killed, waiting greedily for its meal off the creature after men have taken their share of the meat. Herb declared that it had often followed him for hours while he was stealthily tracking a moose, to be in at the death. And now it kept up the din of its unceasing question until he had finished his disagreeable work.

As the party started back to camp, each one weighted with forty pounds or more of meat, Herb carrying a double portion, with the antlers hooked upon his shoulders, they heard the moose-bird still insatiably shrieking "What cheer?" over its meal.

"Say, boys," said the guide, as he stalked along with his heavy load, never blenching, "if you want to get a pair o' moose-antlers, now's your time. I ain't a-going to sell these, but I'll give 'em outright to the first fellow who can learn to call a moose successfully while he's hunting with me. I know what sort of sportsman Cyrus Garst is. He'll go prowling through the woods, starting moose and coolly letting 'em get off without spilling a drop of blood, while he's watching the length of their steps. I b'lieve he'd be a sight prouder of seeing one crunch a root than if he got the finest head in Maine. So here's your chance for a trophy, boys. I guess 'twill be your only one."

"Hurrah! I'm in for this game!" cried Neal.

"I too," said Cyrus.

"I'm in for it with a vengeance!" whooped Dol. "Though I'm blessed if I've a notion what 'calling a moose' means."

"How much have you larned, anyhow, Kid, in the bit o' time you've been alive?" asked the woodsman, with good-humored sarcasm.

"Enough to make my fists talk to anybody who thinks I'm a duffer," answered Dol, squaring his shoulders as if to make the most of himself.

"Good for you, young England!" laughed Cyrus.

Herb turned his eyes, and regarded the juvenile Adolphus with amused criticism.

"Britisher or no Britisher, I'll allow you're a little man," he muttered. "Keep a stiff upper lip, boys; we're not far from camp now."

A word of cheer was needed. Not one of the trio had growled at their load, but the flannel shirts of the two Farrars clung wetly to their bodies. Their breath was coming in hard puffs through spread nostrils. A four-mile tramp through the woods, heavily laden with raw meat, was a novel but not an altogether delightful experience.

However, the smell of moose-steak frying over their camp-fire later on fully compensated them for acting as butcher's boys. When the taste as well as the smell had been enjoyed, the rest which followed by the blazing birch-logs that evening was so full of bliss that each camper felt as if existence had at last drifted to a point of superb content.

Their camp-door stood open for ventilation; and a keen touch of frost, mingling with the night air which entered, made the fragrant warmth delightful.

When supper was ended, and the tin vessels from which it had been eaten, together with all camp utensils, were duly cleaned, Herb seated himself on the middle of the bench, which he called "the deacon's seat," and luxuriously lit his oldest pipe. His brawny hands had performed every duty connected with the meal as deftly and neatly as those of a delicate-fingered woman.

"Well, for downright solid comfort, boys, give me a cosey camp-fire in the wilderness, when a fellow is tired out after a good day's outing. City life can offer nothing to touch it," said Cyrus, as he spread his blankets near the cheerful blaze, and sprawled himself upon them.

Neal and Dol followed his example. The three looked up at their guide, on whose weather-tanned face the fire shed wavering lights, in lazy expectation.

"Now, Herb," said Garst, "we want to think of nothing but moose for the remainder of this trip; so go ahead, and give us some moose-talk to-night. Begin at the beginning, as the children say, and tell us everything you know about the animal."

Herb Heal swung himself to and fro upon his plank seat, drawing his pipe reflectively, and letting its smoke filter through his nostrils, while he prepared to answer.

"Well," he said at last, slowly, "it seems to me that a moose is a troublesome brute to tackle, however you take him. It's plaguy hard for a hunter to get the better of him, and if it's only knowledge you're after, he'll dodge you like a will-o'-the-wisp till you get pretty mixed in your notions about his habits. I guess these English fellows know already that he's the largest animal of the deer tribe, or any other tribe, to be seen on this continent, and as grand game as can be found on any spot of this here earth. I hain't had a chance to chase lions an' tigers; but I've shot grizzlies over in Canada, — and that's scarey work, you better b'lieve! — and I tell you there's no sport that'll bring out the grit and ingenuity that's in a man like moose-hunting. Now, boys, ask me any questions you like, an' I'll try to answer 'em."

"You said something to-day about moose 'crunching twigs,'" began Neal eagerly. "Why, I always had a hazy idea that they fed on moss altogether, which they dug up in the winter with their broad antlers."

"Land o' liberty!" ejaculated the woodsman. "Where on earth do you city men pick up your notions about forest creatures — that's what I'd like to know? A moose can't get its horns to the ground without dropping on its knees; and it can't nibble grass from the ground neither without sprawling out its long legs, — which for an animal of its size are as thin as pipe-stems,— and tumbling in a heap. So I don't credit that yarn about their digging up the moss, even when there's no other food to be had; though I can't say for sure it's not true. In summer moose feed about the ponds and streams, on the long grasses and lily-pads. They're at home in the water, and mighty fine swimmers; so the red men say that they came first from the sea.

"In the fall, and through the winter too, so far as I can make out, they eat the twigs and bark of different trees, such as white birches and poplars. They're powerful fond of moose-wood — that's what you call mountain ash. I guess it tastes to them like pie does to us."

"Well, Dol, I feel that you're twitching all over with some question," said Cyrus, detecting uneasy movements on the part of the younger boy who lay next to him. "What is it, Chick? Out with it!"

"I want to hear about moose-calling," so spoke Dol in heart-eager tones.

The guide swung his body to the music of a jingling laugh.

"Oh; that's it; is it?" he said. "You're stuck on winning those antlers; ain't you,

Dol? Well, calling is the 'moose-hunter's secret,' and it's a secret that he don't want to give away to every one. When a man is a good caller he's kind o' jealous about keeping the trick to himself. But I'll tell you how it's done, anyhow, and give you a lesson sometime. Sakes alive! if you Britishers could only take over a birch-bark trumpet, and give that call in England, you'd make nearly as much fuss as Buffalo Bill did with his cowboys and Injuns. Only 'twould be a one-sided game, for there'd be no moose to answer."

The young Farrars were silent, breathlessly waiting for more. The camp-firelight showed their absorbed faces; it played upon bronzed cheeks, where the ruddy tints of English boyhood had been replaced by a duller, hardier hue. On Neal's upper lip a fine, fair growth had sprouted, which looked white against his sun-tinged skin. As for Cyrus, he had never brought a razor into the woods since that memorable trip when the bear had overhauled his knapsack; so the Bostonian's chin was covered with a thick black stubble.

Neither of the youths, however, was at present giving a thought to his hirsute adornment, about which questionable compliments were frequently bandied. Their minds were full of moose, and their ears alert for the guide's next words.

"P'raps you folks don't know," went on the woodsman, "that there are four ways o' hunting moose. The first and fairest is still-hunting 'em in the woods, which means following their signs, and getting a shot in any way you can, *if* you can. But that's a stiff 'if' to a hunter. Nine times out o' ten a moose will baffle him and get off unhurt, even when a man has tracked him for days, camping on his trail o' nights. The snapping of a twig not the size of my little finger, or one tramping step, and the moose'll take warning. He'll light out o' the way as silently as a red man in moccasins, and the hunter won't even know he's gone.

"The second way is night-hunting, going after 'em in a canoe with a jack-light; same thing as jacking for deer. I guess you've tried that, so you'll know what it's like — skeery kind o' work."

Neal nodded an eloquent assent, and Herb went on: —

"The third method is a dog's trick. It's following 'em on snowshoes over deep snow. I've tried that once, and I'm blamed if I'll ever try it again. It's butchery, not sport. The crust of snow will be strong enough for a man to run on, but it can't support the heavy moose. The creature'll go smashing through it and struggling out, until its slim legs are a sight to see for cuts and blood. Soon it gets blowed, and can stumble no farther. Then the hunter finishes it with an axe."

Disgust thickened the voices of the listening three, as with one accord they raised an outcry against this cruel way of butchering a game animal, without giving it a single chance for its life. When their indignation had subsided, the hunter went on to describe the fourth and last method of entrapping moose — the calling in which Dol was so interested.

"P'raps you won't think this is fair hunting either," he said; "for it's a trick, and I'll allow that there's times when it seems a pretty mean game. Anyhow, I'd rather kill one moose by still-hunting than six by calling. But if you want to try work that'll make your blood race through your body like a torrent one minute, and turn you as cold as if your sweat was ice-water the next, you go in for moose-calling. I guess you know all about the matter, Cyrus; but as these Britishers do not, I'll try and explain it to 'em.

"Early in September the moose come up from the low, swampy lands where they have spent the summer alone, and begin to pair. Then the bull-moose, as we call the male, which is generally the most wide-awake of forest creatures, loses some of his big caution, an' goes roaming through the woods, looking for a mate. This is the time for fooling him. The hunter makes a horn out o' birch-bark, somewheres about eighteen inches long, through which he mimics the call of the cow-moose, to coax the bull within reach of his rifle-shots."

"What is the call like?" asked Neal, his heart thumping while he remembered that strange noise which had marked a new era in his experience of sounds, as he listened to it at midnight by Squaw Pond.

"Sho! a man might keep jawing till crack o' doom, and not give you any idea of it without you heard it," answered Herb Heal, the dare-all moose-hunter. "The noise begins sort o' gently, like the lowing of a tame cow. It seems, if you're listening to it, to come rolling — rolling — along the ground. Then it rises in pitch, and gets impatient and lonely and wild-like, till you think it fills the air above you, when it sinks again and dies away in a queer, quavery sound that ain't a sigh, nor a groan, nor a grunt, but all three together.

"The call is mostly repeated three times; and the third time it ends with a mad roar as if the lady-moose was saying to her mate, '*Come* now, or stay away altogether!'"

"Joe Flint was right, then!" exclaimed Neal, in high excitement. "That's the very noise I heard in the woods near Squaw Pond, on the night when we were jacking for deer, and our canoe capsized."

"P'raps it was," answered Herb, "though the woods near Squaw Pond ain't much good for moose now. They're too full of hunters. Still, you might have heard the cow-moose herself calling, or some man who had come across the tracks of a bull imitating her."

"But if the bull has such sharp ears, can't he tell the real call from the sham one?" asked Dol.

"Lots of times he can. But if the hunter is an old woodsman and a clever caller, he'll generally fool the animal, unless he makes some awkward noise that isn't in the game, or else the moose gets his scent on the breeze. One whiff of a man will send the creature off like a wind-gust, and earthquakes wouldn't stop him. And though he sneaks away so silently when he *hears* anything suspicious, yet when he *smells* danger he'll go through the forest at a thundering rush, making as much noise as a demented fire-brigade."

"Good gracious!" ejaculated Neal and Dol together.

"Is the moose ever dangerous, Herb?" asked the former.

"I guess he is pretty often. Sometimes a bull-moose will turn on a hunter, and make at him full tilt, if he's in danger or finds himself tricked. And he'll always fight like fury to protect his mate from any enemy. The bulls have awful big duels between themselves occasionally. When they're real mad, they don't stop for a few wounds. They prod each other with their terrible brow antlers till one or the other of 'em is stretched dead. If a moose ever charges you, boys, take my advice, and don't try to face him with your rifles. Half a dozen shots mightn't stop him. Make for the nearest tree, and climb for your lives. Fire down on him then, if you can. But once let him get a kick at you with his forefeet, and one thing is sure — *you'll* never kick again. Are you tired of moose-talk yet?"

"Not by a jugful!" answered Cyrus, laughing. "But tell us, Herb, how are we to proceed to get a sight of this 'Jabberwock' alive?"

"If to-morrow night happens to be dead calm, I might try to call one up," answered the guide. "There's a pretty good calling-place near the south end of the lake. As this is the height of the season, we might get an answer there. We'll try it, anyhow, if you're willing."

"Willing! I should say we are!" answered Garst. "You're our captain now, Herb, and it's a case of 'Follow my leader!' Take us anywhere you like, through jungles or mud-swamps. We won't kick at hardships if we can only get a good look at his mooseship. Up to the present, except for that one moonlight peep, he has always dodged me like a phantom."

"Are you going to be satisfied with a look?" The guide's eyes narrowed into two long slits, on which the firelight quivered, as he gazed quizzically down upon Cyrus. "If the moose comes within reach of our shots, ain't anybody going to pump lead into him? Or is he to get off again scot-free? I've got my moose for this season, and I darsn't send my bullets through the law by dropping another, so I can't do the shooting."

"My friends can please themselves," said the Bostonian, glancing at the English lads. "For my own part I'll be better pleased if Mr. Moose manages to keep a whole skin. Our grand game is getting scarce enough; I don't want to lessen it. I once saw the last persecuted deer in a county, after it had been badgered and wounded by men and dogs, limp off to die alone in its native haunts. The sight cured me of bloodthirst."

"I guess 'twould be enough to cure any man," responded Herb. "And we don't want meat, so this time we won't shoot our moose after we've tricked him. Good land! I wouldn't like any fellow to imitate the call of my best girl, that he might put a bullet through me. Come, boys, it's pretty late; let's fix our fire, and turn in."

## CHAPTER XVI.

#### MOOSE—CALLING.

NOTHING was talked about among the campers on the following day but the forthcoming sport of the evening — moose-calling.

Herb Heal had decided that his call should be given from the water, his "good calling-place" being an alder-fringed logon at the loneliest extremity of the lake.

During the afternoon he took Neal and Dol with him into a grove of poplars and birches which bordered one end of the clearing, leaving Cyrus lounging by the camp-fire. Here the woodsman began the exciting work of preparing his birch-bark horn, that primitive but potent trumpet through which he would sigh, groan, grunt, and roar, imitating each varying mood of the cow-moose. To her call he had often listened as he lay for hours on a mossy bed in the far depths of the forest, learning to interpret the language of every woodland creature.

Unsheathing his hunting-knife, and selecting a sound white-birch tree, Herb carefully removed from it a piece of bark about eighteen inches in length and six in width. This he carefully trimmed, and rolled into a horn as a child would twist paper into a cornucopia package for sweets, tying it with the twine-like roots of the ground juniper. The tapering end of the trumpet, which would be applied to the caller's lips, measured about one inch across; its mouth measured five.

Returning to camp, Herb dipped the horn in warm water and then let it dry, saying that this would produce a mellow ring. He stoutly refused all appeals from the boys to give them a few illustrations of moose-calling there and then, with a lesson in the art, declaring that it would spoil the night's sport, and that they must first hear the call amid proper surroundings. From time to time he impressed upon them that they were going to engage in an expedition which required absolute silence and clever stratagem to make it successful. He vowed to wreak a woodsman's vengeance on any fellow who balked it by shaking the boat, or by moving body or rifle so as to make a noise.

A light, humming breeze had been blowing all day; but as the afternoon waned, it died down. The evening proved clear, chilly, and still.

"Is this a likely night for calling, Herb?" asked Cyrus anxiously, taking a survey of sky and lake from the camp-door about an hour before the start.

"Fine," answered Herb with satisfaction. "Guess we'll get an answer sure, if there's a moose within hearing. There ain't a puff of wind to carry our scent, and give the trick away. But rig yourselves up in all the clothing you've got, boys; the cold, while we're waiting, may be more than you bargain for."

The guide had a light boat on the lake, moored below the camp. At six o'clock he seated himself therein, taking the oars in his brawny hands. Cyrus and Neal took their places in the stern; while Dol disposed of himself snugly in the bow, right under a jack-lamp which Herb had carefully trimmed and lit. But he had closed its sliding door, which, being padded with buckskin, could be opened and shut without a sound, so that not a ray of light at present escaped.

"Moose won't stand to watch a jack as deer do," he said. "'Twill only scare 'em off. They're a heap too cute to be taken in by an onnatural big star floating over the water. But 'taint the lucky side of the moon for us. She'll rise late, and her light 'll be so feeble that it wouldn't show us an elephant clearly if he was under our noses. So if I succeed in coaxing a bull to the brink of the water, I'll open the jack, and flash our light on him. He'll bolt the next minute as quick as greased lightning on skates; but if you only get a short sight of him, I promise that 'twill be one you'll remember."

"And if he should take a notion to come for us?" said Cyrus.

"He won't, if we don't fire. The boat will be lying among the black shadows, snug in by the bank, and he'll see nothing but the dazzling light. But you fellows must keep still as death. Off we go now, boys, and mum's the word!"

This was almost the last sentence spoken. Not a syllable moved the lips of any one of the four, as the boat glided away from camp towards the south end of the lake, the oars making scarcely a sound as Herb handled them. By and by he ceased rowing for an instant, took his pipe from his mouth, knocked out its ashes, and put it in his pocket with a wise look at his companions, murmuring, "Don't want no tobacco incense floating around!"

At the same time, from a distant ridge upon the eastern shore, covered with evergreens which stood out like dark steeples against the evening sky, came a faint, dull noise, as if some belated woodsman was driving a blunt axe against a tree. The sound itself would scarcely have awakened a hope of anything unusual in the minds of the inexperienced; but, combined with the guide's aspect as he pocketed his pipe, it made Cyrus and his comrades sit suddenly erect, listening as if ears were the only organs they possessed.

The queer, dull noise was once repeated. Then again there was silence almost absolute, Herb's oars moving with the softest swish imaginable, as the boat skimmed along the lonely, curved bay which he had chosen for a calling-place. It came to a stop amid shadows so dense and black that they seemed almost tangible, close to a bank fringed with overhanging bushes, having a background of evergreens. These last, in the fast-gathering darkness, looked like a sable array of mourners in whose ranks a pale ghost or two mingled, the spectres being slim white-birch trees.

The opposite bank presented a similar scene.

It was amid such surroundings that Neal Farrar heard for the second time in his life the weird sound of the moose-hunter's call. He was a strong, well-balanced young fellow; yet here again he knew the sensation as if needles were pricking him all over, which he had felt once before in these wilds, while his heart seemed to be performing athletic sports in his body.

Cyrus and Dol confessed afterwards that they were "all shivers and goose-flesh" as the call rose upon the night air.

After he had shipped his oars, and laid them down, Herb Heal noiselessly turned his body to face the bow, and took up the birch-bark horn which lay beside him. He breathed into it anxiously once or twice, then paused, drew in all the air which his big lungs could contain, put the trumpet again to his lips with its mouth pointing downward, and began his summons.

The first part of the call lasted half a minute, or so, without a break. During its execution the hunter moved his neck and shoulders first to the left, then to the right, and slowly raised the horn above his head, the rolling, plaintive sounds with which he commenced gathering power and pitch with the ascending motion. As the birch trumpet pointed straight upward, they seemed to sweep aloft in a surging crescendo, and boom among the tree-tops.

Carrying his head again to the left and right, Herb gradually lowered the horn until it was once more pointed towards the bottom of the boat, having in its movements described in the air a big figure of eight. The call sank with it, and died away in a lonely, sighing, quavering grunt.

Two seconds' pause, two slow, great throbs of the boys' hearts, so loud that they threatened to burst the stillness.

Then the call began again, low and grumbling. Again it rose, swelled, quavered, and sank, full of lonely longing.

A third time it surged up, and ended abruptly in a wild, ear-splitting roar, which struck the tops of distant hills, and rolled off in thunder-like echoes among them.

Silence followed. Not a gasp came from Herb after his efforts. Cyrus and the Farrars tried to still their heaving chests, while each quick breath was an expectation.

An answer! Surely it was an answer! The boys never doubted it; though the responding sound they caught was only a repetition of that far-away chopping noise, which resembled the heavy thud of an axe against wood. This came nearer — nearer. It was followed once by a sort of short, sharp bark.

Then the motionless occupants of the boat heard random, guttural grunts, a smashing of dead branches, crashing of undergrowth, and the proud ring of mighty antlers against the trees. The lord of the forest, a big bull-moose, was tearing recklessly through the woods towards the lake, in answer to the call of his imaginary mate.

To say that the hearts of our trio were performing gymnastic feats during these awfully silent minutes of waiting, is to say little. All the repressed motion of their bodies seemed concentrated in these organs, which raced, leaped, stopped short, and pounded, vibrating to such questions as: —

"Will he come? Where shall we first see him? How near is he now? Does he suspect the trick? Will he give us the slip after all? — *Has he gone?*"

For of a sudden dead stillness reigned in the forest. No more trampling, grunting, and knocking of antlers. The spirits of the three sank to zero. Their breathing became thick. The blood, which a moment before had played like wildfire in their veins, now stirred sluggishly as if it was freezing. Disappointment, blank and bitter, shivered through them from neck to foot.

So passed quarter of an hour. A filmy mist rose from the surface of the water, and drifted by their faces like the brushing of cold wings. For lack of motion hand and feet felt numb. Mid the pitch-black shadows, snug in by the bank, no man could see the face of his fellow, though the trio would have given a fortune to read their guide's. Not a word was spoken. Once, when a deep breath of impatience escaped him, Neal heard the folds of his coat rub each other, and clenched his teeth to stop an exclamation at the sound, which he had never noticed before.

Nearly twenty minutes had elapsed since the last noise had been heard in the woods, when Herb took up the horn which he had laid down, and put it to his mouth. Again the call rolled up. It was neither loud nor long this time, ending with a quick, short roar.

As it ceased the guide plunged his arm into the water and slowly withdrew it, letting drops dribble from his fingers.

The novices could only suspect that this manœuvre was another lure for the bull-moose, if he chanced to be still within hearing. Its success took their breath away.

The wary bull which had answered, having doubtless harbored a suspicion that all was not exactly right with the first call, had halted in his on-coming rush, with head upreared, and nostrils spread, trying to catch any taint in the air which might warn him of danger. But in the dead calm the heavy evergreens stirred not; no whiff reached him. The second call upset his prudence. Then he heard that splash and dribble in the water, and imagined that his impatient mate was dipping her nose into the lake for a cool drink.

A snort! A bellowing challenge quite indescribable! On he came again with a thundering rush!

Bushes were thrashed and spurned by his sharp hoofs. Branches snapped. Trees echoed as his antlers struck them.

A musk-rat leaped from the bank ahead, and dived to reach his hole in the bank. Under cover of the noisy splash which the little creature made, one whisper was hissed by Herb's tongue into the ears of his comrades. It was: —

"Gee whittaker! he's a big one! Listen to them shovels against the trees!"

A minute later, with a deep gulp of intense excitement, and a general racket as if an engine had broken loose from brakes and checks, and was carrying all before it, the monarch of the woods crashed through the alders and halted, with his hoofs in the water, scarcely thirty yards from where the boat lay in shadow.

This was a supreme moment for our travellers. Leaning forward, fearful lest their heart-beats should betray them, they could barely distinguish the outlines of the moose, as he stood with his enormous nose high in air, giving vent to deep gulps and grunts, and looking to right and left in bewilderment for that cow which he had heard calling.

For fully five minutes he stood thus, badly

puzzled, now and again stamping a hoof, and scattering spray in rising wrath. Then Herb bent forward, shot out a long arm, and silently opened the jack.

Meteor-like its silver light flashed forth, to reveal a sight which could never be wiped from the memories of the beholders, though it affected each of them differently.

Herb Heal involuntarily gripped the loaded rifle which lay beside him, — he was too wary a woodsman to be unprepared for emergencies; but he did not cock it, for he remembered the law, and the bargain which he had made about to-night.

Cyrus's eyes gleamed like fires in a face pale from eagerness, as he strove in a minute of time to take in every feature of the monster before him, from hoof to horn.

Neal sat as if paralyzed.

Dol — well, Dol lost his head a bit. A deep, throaty gulp, which was a weak reproduction of the sound made by the moose, as if the boy and the animal were sharing the same throes of excitement, burst from him. There was a rattle and struggle of his vocal organs, which in another second would have become a shout, had not Herb's masterful left hand gripped him. Its touch held in check the speech which Dol could no longer control.

The moose was a big one, "about as big as they grow," as the guide afterwards declared. Under the jack-light he looked a regular behemoth. He must have been over seven feet high at the shoulders, for he was taller than the tallest horse the boys had ever seen. His black mane bristled. His antlers were thrown back. His great nose, with its dilated nostrils, looked as if it were drinking in every scent of the night world. His eyes had a green glare in them, as for ten seconds he gazed at the strange light which had suddenly burst into view, its silver radiance so dazzling him that he saw not the screened boat beneath.

At the rash noise which Dol made his ears twitched. He splashed a step forward as if to investigate matters, seeing which, Herb held his Winchester in readiness to fly to his shoulder at a moment's notice. But the moose evidently regarded the jack-lamp as a supernatural, terrible phenomenon. He shrank from it as man might shrink beneath a flaming heaven.

With one more despairing look right and left for that phantom cow which had deluded him, he wheeled around, and crashed back into the forest, tearing away more rapidly than he came.

"He's off now, and Heaven knows when he'll stop!" said Herb, breaking the weird spell of silence. "Not till he reaches some lair where nary a creature could follow him. Well, boys, you've seen the grandest game on this continent, the king o' the woods. What do you think of him?"

All tongues were loosened together. There was a general shifting of cramped bodies, accompanied by a gust of exclamations.

"He was a monster!"

"He was a behemoth!"

"Oh! but you're a conjurer, Herb. How on earth did you give such a fetching call?"

"I could never have believed that those sounds came from a human throat and a birch-bark horn, if I hadn't been sitting in the boat with you!"

When there was a break in the excited chorus, Herb, without answering the compliments to his calling powers, asked quietly,—

"Didn't you think we'd lost him, boys, when he stopped short in the middle of his rush, and you heard nothing?"

"We just did," answered Cyrus. "That

was the longest half-hour I ever put in. What made him do it?"

"I guess he was kind o' criticising my music," said the guide, laughing. "Mebbe I got in a grunt or two that wasn't natural, and the old boy wasn't satisfied with his sweetheart's voice. He was sniffing the air, and waiting to hear more. But 'twasn't more 'n twenty minutes before I gave the second call, though no doubt it seemed longer to you. A man must be in good training to get the better of a moose's ears and nose."

"I'm going to get the better of them before I leave these woods!" cried Dol, who was still puffing and gasping with intense excitement. "I'll learn to call up a moose, if I crack my windpipe in doing it."

"Hurrah for the Boy Moose-Caller!" jeered Cyrus, with a teasing laugh, which Neal echoed.

But Herb Heal, who had from the beginning regarded "the kid of the camp" with favor, suddenly became his champion.

"Don't let 'em down you, Dol," he said. "I hate to hear a youngster, or a man, 'talk fire,' as the Injuns say, which means *brag*, if he's a coward or a chump; but I guess you ain't either. Here we are at camp, boys! I tell you the home-camp is a pleasant sort of place, after you've been out moose-calling!"

Thereupon ensued loud cheers for the home-camp, the boys feeling that they were letting off steam, and atoning for that long spell of silence, which had been a positive hardship. In the midst of an echoing hubbub the boat was hauled up and moored, and the party reached their log shelter.

## CHAPTER XVII.

#### HERB'S YARNS.

THE following day was spent by our trio in exploring the woods near Millinokett Lake, in listening to more moose-talk, and in attempting the trick of calling. Herb gave them many persistent lessons, making the sounds which he had made on the preceding night, with and without the horn, and patiently explaining the varied language of grunts, groans, sighs, and roars in which the cow-moose indulges.

Perhaps the woodsman expended extra pains on the teaching of his youngest pupil, whom he had championed. And certainly Dol's own talent for mimicry came to his aid. No matter to what cause the success was due, each one allowed that Dol made a brilliant attempt to get hold of "the moose-hunter's secret," and give a natural call.

The boy had been a genius at imitating the voices of English birds and animals; many a trick had he played on his schoolfellows with his carols and howls. And his proficiency in this line was a good foundation on which to work.

"You'll get there, boy," said Herb, surveying him with approval, as he stood outside the camp-door with the moose-horn to his lips. "Make believe that there's a moose on the opposite shore of the lake now, and give the whole call, from start to finish."

Whereupon Dol slowly carried his head to left and right, as he had seen the guide do on the previous night, raising and lowering the horn until it had described an enormous figure of eight in the air, while he groaned, sighed, rasped, and bellowed with a plaintive intensity of expression, which caused his brother and his friend to shriek with laughter.

"You'll get there, Kid," repeated the woodsman, with a great triumphant guffaw. "You'll be able to give a fetching call sooner than either of the others. But be careful how you use the trick, or you'll be having the breath kicked out of you some day by a moose's fore-feet."

For days afterwards, the birch-bark horn was rarely out of Dol Farrar's hands. The boy was so entranced with the new musical art he was mastering, which would be a means of communication between him and the behemoth of the woods, that he haunted the edges of the forest about the clearing, keeping aloof from his brother and friend, practising unceasingly, sometimes under Herb's supervision, sometimes alone. He learned to imitate every sound which the guide made, working in touching quavers and inflections that must tug at the heart-strings of any listening moose. He learned to give the call, squatting Indian fashion, in a very uncomfortable position, behind a screen of bushes. He learned to copy, not the cow's summons alone, but the bull's short challenge too; and to rasp his horn against a tree, in imitation of a moose polishing its antlers for battle.

And now, for the first time, Dol Farrar of Manchester regarded his education as complete. He was prouder of this forest accomplishment, picked up in the wilds, than of all triumphs over problems and 'ologies at his

English school. He had not been a laggard in study, either.

But the finishing of Dol's education had one bad result. If there happened to be another moose travelling through the adjacent forests, he evidently thought that all this random calling was too much of a good thing, had his suspicions aroused, and took himself off to wilder solitudes. Though the guide tried his powers in persuasive summons every night at various calling-places, he could not again succeed in getting an answer.

At last, on a certain evening, after supper, a solemn camp-council was held around an inspiring fire, and Herb Heal suggested that if his party were really bent on seeing a moose again, before they turned their faces homeward, they had better rise early the following morning, shoulder their knapsacks, and set out to do a few days' hunting amid the dense woods near the base of Katahdin.

"I killed the biggest bull-moose I ever saw, on Togue Ponds, in that region," said the guide meditatively; "and I got him in a queer way. I b'lieve I promised to tell you that yarn."

"Of course you did!"

"Let's have it!"

"Go ahead, Herb! Don't shorten it!"

Thus encouraged by the eager three, the woodsman began: —

"It is five years now, boys, since I spent a fall and winter trapping in them woods we were speaking of — I and another fellow. We had two home-camps, which were our headquarters, snug log shelters, one on Togue Ponds, the other on the side of Katahdin. As sure as ever the sun went down on a Saturday night, we two trappers met at one or other of these home-camps; though during the week we were mostly apart. For we had several lines of traps, which covered big distances in various directions; and on Monday morning I used to start one way, and my chum another, to visit these. Generally it took us five or six days to make the rounds of them. While we were on our travels we'd sleep with a blanket round us, under any shelter we could rig up, — a few spruce-boughs or a bark hut. When the snow came, we were forced to shorten our trips, so as to reach one of the home-camps each night.

"Well, it was early in the season, one fine fall evening, that I was crossing Togue Ponds in a canoe. I had been away on the tramp for a'most a week; and though I had a rifle and axe with me, I had nary an ounce of ammunition left. All of a sudden I caught sight of a moose, feeding on some lily-roots in deep water. Jest at first I was a bit doubtful whether it was a moose or not; for the creature's head was under, and I could only see his shoulders. I stopped paddling. I tried to stop breathing. Next, I felt like jumping out of my skin; for, with a big splash, up come a pair of antlers a good five feet across, dripping with water, and a'most covered with green roots and stems, which dangled from 'em.

"Good land! 'twas a queer sight. 'Herb Heal,' thinks I, 'now's your chance! If you can only manage to nab that moose-head, you'll get two hundred dollars for it at Greenville, sure!' And mighty few cents I had jest then.

"I could a'most have cried over my tough luck in not having one dose of lead left. But the bull's back was towards me. The water filled his ears and nose, so that he couldn't hear or smell. And he was having a splendid tuck-in. It was big sport to hear him crunch those lily-roots."

"I should think it was!" burst out Cyrus enviously. "But did you have the heart to kill him in cold blood, in the middle of his meal?"

"I did. I guess I wouldn't do it now; anyhow, not unless I was very badly off for food. But I had an old mother living at Greenville that time," — here there was the least possible tremble in the woodsman's voice, — "and while I paddled alongside the moose, without making a sound, I was thinking that the price I'd be sure to get from some city swell for the head would come in handy to make her comfortable. The creature never suspicioned danger till I was close to him, and had my axe lifted, ready to strike. Then up came his head. Out went his forefeet. Over spun the canoe. There was as big a commotion as if a whale was there.

"I managed to keep behind the brute so as to dodge his kicks; and gripping the axe in one hand, I dug the other into his long hair. He was mad scared. He started to swim for the opposite shore, which was about half a mile distant, with me in tow, snorting like a locomotive. As his feet touched ground near the bank, I jumped upon his back. With one blow of the axe I split his spine. Perhaps you'll think that was awful cruel, but it wasn't done for the glory of killing."

"And what became of the head? Did you sell it?" asked Dol, who was, as usual, the first to break a breathless silence.

There was no reply. Herb feigned not to hear.

"Did you get two hundred dollars for the head?" questioned the impetuous youngster again, in a higher key, his curiosity swelling.

"I didn't. It was stole."

The answer was a growl, like the growl of a hurt animal whose sore has been touched. The tone of it was so different from the woodsman's generally strong, happy-go-lucky manner of speech, that Dol blenched as if he had been struck.

"Who stole it?" he gasped, after a minute, scarcely knowing that he spoke aloud.

Unnoticed in the firelight, Cyrus clapped a strong hand over the boy's mouth, to stifle further questions.

"Keep still!" he whispered.

But Herb, who was, as usual, perched upon the "deacon's seat," leaned forward, with a laugh which was more than half a snarl.

"Who stole it?" he echoed. "Why, the other fellow — my chum; the man whom I carried for a mile on my back, through a snow-heaped forest, the first time I saw him,

The Camp on Millinokett Lake.

when I had lugged him out of a heavy drift. *He* stole it, Kid, and a'most everything I owned with it."

With a savage kick of his moccasined foot, the woodsman suddenly assaulted a blazing log. It sent a shower of sparks aloft, and caused a bright flame to shoot, rocket-like, from the heart of the fire, which showed the guide's face. His fine eyes reminded Cyrus of Millinokett Lake when a thunder-storm broke over it. Their gray was dark and troubled; the black pupils seemed to shrink, as if a tempest beat on them; fierce flashes of light played through them.

Muttering a half-smothered oath, Herb flung himself off his bench, stamped across the cabin to the open camp-door, and passed into the darkness outside.

The boys, who had been stretched out in comfortable positions, drew themselves bolt upright, and sat aghast. They stared towards the camp-door, murmuring disjointedly. Into the mind of each flashed a remembrance of some story which Doctor Phil had told about a thieving partner who once robbed Herb Heal.

"You've stirred up more than you bargained for, Dol," said Cyrus. "I wish to goodness you hadn't been so smart with your questions."

But the words were scarcely spoken when the guide was again in their midst, with a smile on his lips.

"It's best to let sleeping dogs lie, young one," he said, looking down reassuringly on Dol, who was feeling dumfounded. "I guess you all think I'm an awful bearish fellow. But if you had lived the lonely life of a trapper, tramping each day through the dark woods till you were leg-weary, visiting your steel traps and deadfalls, all to get a few furs and make a few dollars; and turned up at camp one evening to find that your partner had skipped with every skin you had procured, I reckon 'twould take you a plaguy long time to get over it."

"I'm pretty sure it would, old man," said Cyrus.

"And I minded the loss of the furs a sight less than I minded losing that moose-head," continued Herb, taking his perch again upon the "deacon's seat." "The hound took 'em all. Every woodsman in Maine was riled about it at the time, and turned out to ketch him; but he gave 'em the slip. Now, boys, I've got to feeling pretty chummy with you.

Cyrus is an old friend; and, to speak plain, I like you Britishers. I don't want you to think that I bust up your fun to-night for nothing. I'll tell you the whole yarn if you want to hear it."

The looks of the trio were sufficient assent.

"All right, boys. Here goes! Since I was a kid in Maine woods I've worked at a'most everything that a woodsman can do. Six year ago I was a 'barker' in a lumber-camp on the Kennebec River. A 'barker' is a man who jumps onto a big tree after a chopper has felled it, and strips the bark off with his axe, so that the trunk can be easily hauled over the snow. Well, it's pretty hard labor, is lumbering. But our camp always got Sunday for rest.

"Well, I was prowling about in the woods by myself one Sunday afternoon, when an awful snow-storm come on, a big blizzard which staggered the stripped trees like as if 'twould tumble 'em all down, and end our work for us. I was bolting for camp as fast as I was able, when I tripped over something which was a'most covered over in a heavy drift. 'Great Scott!' says I, 'it's a man!' And 'twas too. He was near dead. I hauled him out, and set him on his legs; but he couldn't walk. So I threw him across my shoulders, same way as I carry a deer. He didn't weigh near as much as a good buck, for he was little more'n a kid and awful lean. But 'twas dreadful travelling, with the snow half blinding and burying you. I was plumb blowed when I struck the camp, and pitched in head foremost.

"For an hour we worked over that stranger to bring him round, and we succeeded. We saw at once that he was a half-breed. When he could use his tongue, he told us that his father was a settler, and his mother a Penobscot Indian. He was sick for a spell and wild-like, then he talked a lot of Indian jargon; but when he got back his senses, he spoke English fust-rate. Chris Kemp he said was his name. And from the start the lumbermen nicknamed him 'Cross-eyed Chris;' for his eyes, which were black as blackberries, had a queer squint in 'em.

"Well, in spite of the squint, I took to Chris, and he to me. And the following year, when I decided to give up lumbering, and take to trapping fur-bearing animals in the woods near Katahdin, he joined me. We swore to be chums, to stick to each other through thick and thin, to share all we got;

and he made one of his outlandish Indian signs to strengthen the oath. A fine way he kept it too!

"Now, if I'm too long-winded, boys, say so; and I'll hurry up."

"No, no! Tell us everything."

"Spin it out as long as you can."

"We don't mind listening half the night. Go ahead!"

At this gust of protest Herb smiled, though rather soberly, and went ahead as he was bidden.

"We made camp together — him and me. We had two home-camps where I told you, and met at the end of each week, bringing the skins we had taken, which we stored in one of 'em. We got along together swimmingly for a bit. But Chris had a weakness which I had found out long before. I guess he took it from his mother's people. Give him one drink of whiskey, and it stirred up all the mud that was in him. There's mud in every man, I s'pose; and there's nothing like liquor for bringing it to the surface. A gulp of fire-water changed Chris from an honest, right-hearted fellow to a crazy devil. This had set the lumbermen against him. But I hoped that in the lonely woods where we trapped he wouldn't get a chance to see the stuff. He did, though, and when I wasn't there to make a fight against his swallowing it.

"It happened that one week he got back to our camp on Togue Ponds, — where most of our stuff was stored, and where I kept that moose-head, waiting for a chance to take it down to Greenville, — a day or two sooner'n me. And the worst luck that ever attended either of us brought a stranger to the camp at the same time, to shelter for a night. He was an explorer, a city swell; and I guess he didn't know much about Injuns or half-breeds, for he gave Chris a little bottle of fiery whiskey as a parting present. The man told me about it afterwards, and that he was kind o' scared when the boy — for he wasn't much more — swallowed it with two gulps, and then followed him into the woods, howling, capering, and offering to sell him my grand moose-head, and all the furs we had, for another drink of the burning stuff. I guess that stranger felt pretty sick over the mischief he had done. He refused to buy 'em. But when I got back to camp next day, to find the skins gone, antlers gone, Chris gone; when I ran across the traveller and ferreted out his story,—I knew, as well as if I seen it, that my partner had skipped with all my belongings, to sell 'em or trade 'em at some settlement for more liquor. We had a couple of big birch canoes,—one of 'em was missing too,—and a river being near, the thing could be easy managed.

"I'll allow that I raged tremendous. The losses were bad; but to be robbed by your own chum, the man you had saved and stuck to, the only being you had said a word to for months, was sickening. I swore I'd shoot the hound if I found him. I spread the news at every camp and farm-settlement through the forest country, and we had a rousing hunt after the fellow; but he gave us the slip, though I heard of him afterwards at a distant town, where he sold the furs."

"I suppose he left the State," said Cyrus.

"I guess he did. But for a big while I used to think he'd come back to our camp some day, and let me have it out with him; for he wasn't a coward, and we had been fast chums."

"And he didn't?"

"Not as I know of. The next year I gave up trapping, which was an awful cruel as well as a lonely business, and took to moose-hunting and guiding. I haven't been anear the old camps for ages."

"Perhaps you will come across him again some day," suggested Dol, with unusual timidity.

"P'raps so, Kid. And, faith, when I think of that, it seems as if there were two creatures inside o' me fighting tooth and claw. One is all for hammering him to a jelly. The other is sort o' pitiful, and says, 'Mebbe 'twasn't out-an'-out his fault.' Which of them two'll get the best of it, if ever I'm face to face with Cross-eyed Chris, I dunno."

Cyrus Garst rose suddenly. He kicked the camp-fire to make a blaze, then looked the woodsman fair in the eyes.

"I know, Herb," he said; "the spirit of mercy will conquer."

"Glad you think so!" answered Herb. "But I ain't so sure. Sho! boys, I've kept you up till near midnight with my yarns. We must go to roost quick, or you'll never be fit to light out for Katahdin to-morrow."

## CHAPTER XVIII.

### TO LONELIER WILDS.

BEFORE daybreak next morning Herb Heal was astir. Apparently even a short night's sleep had driven from him all disturbing memories. He whistled and hummed softly, like the strong, hopeful fellow he was, controlling his notes so that they should not awaken his companions, while he hauled out and overlooked the canvas for a tent, to see if it was sound. Next he surveyed the camp-stores, and put up a supply of flour, pork, and coffee in a canvas bag, enough for four persons to subsist upon with economy during an excursion of six or seven days. For he knew that his employers would follow his suggestion, and be eager to start for the woods near Katahdin soon after they got their eyes open.

He had been doing his work with a candle held in his brown fingers; but as dawn-light began to enter the cabin, he quenched its dingy, yellow flicker, opened the camp-door, and surveyed the morning sky.

"It'll be a good day to start out, I guess," he muttered. "Let's see, what time is it?"

The stars had not yet paled, and Herb forthwith fell to studying them; for they were his jewelled time-piece, by which he could tell the hour so long as they shone. Watch he had none.

While he gazed aloft at the glinting specks, he unconsciously began to croon, in a powerful bass voice, with deep gutturals, some words which certainly weren't woodsman's English.

> "*N'loan pes-saus, mok glint ont-aven,
> Glint ont-aven, nosh morgun.*"

"What on earth is that outlandish thing you're singing, Herb?" roared Neal Farrar from the bunk, awakened by the sounds. "Give us that stave again — do!"

The guide started. He had scarcely been aware of what he was humming, and his laugh was a trifle disconcerted.

"So you're waking up, are ye?" he said. "'Tain't time to be stirring yet; I ought to be kicked for making such a row."

"But what's that you were singing?" reiterated Neal. "The words weren't English, and they had a fine sort of roll."

"They're Injun," was the answer. "I guess 'twas all the talking I done last night that brung 'em into my head. I picked 'em up from that fellow I was telling you about. He'd start crooning 'em whenever he looked at the stars to find out the hour."

"Are they about the stars?"

"I guess so. A city man, who had studied the redskins' language a lot, told me they meant:—

> 'We are the stars which sing,
> We sing with our light.'"[1]

Then Herb chanted the two lines again in the original tongue.

"There was quite a lot more," he said; "but I can't remember it. I learned some queer jargon from Chris, and how to make most of the signs belonging to the Indian sign-talk. The fellow had more of his mother than his father in him. I guess I'd better give over jabbering, and cook our breakfast."

[1] Mr. Leland's translation.

It was evident that Herb did not want to dwell upon his reminiscences. And Neal had tact enough to swallow his burning curiosity about all things Indian. He asked no more questions, but rolled off the fir-boughs, and dressed himself.

Cyrus and Dol sprang up too. All three were soon busy helping forward preparations for the start. They packed their knapsacks with a few necessaries; and after a hearty breakfast had been eaten, — their last meal off moose-steaks for a while, as Herb informed them he "could not carry any fresh meat along," — the guide's voice was heard shouting: —

"Ready, are ye, boys? Got all yer traps? Here, Cyrus, jest strap this pack-basket on my shoulders. Now we're off!"

The pack contained the tent, the camp-kettle, and frying-pan, together with the aforementioned provisions, a good axe, etc. It was an uncomfortable load, even for a woodsman's shoulders. But Herb strode ahead with it jauntily. And many times during that first day's tramp of a dozen miles, his comrades — as they trudged through rugged places after him, spots where it was hard to keep one's perpendicular, and feet sometimes showed a sudden inclination to start for the sky—threw envious glances at his tall figure, "straight as an Indian arrow," his powerful limbs, and unerring step. Even the horny, capable hands came in for a share of the admiration.

"I guess anything that got into your grip, Herb, would find it hard to get out again without your will," said Cyrus, studying the knotted fists which held the straps of the pack-basket.

"Mebbe so," answered the guide frankly. "I've a sort of a trick of holding on to things once I've got 'em. P'raps that was why I didn't let go of Chris in that big blizzard 'till I landed him at camp. But I hope "— here Herb's shoulders shook with heaving laughter, and the cooking utensils in his pack jingled an accompaniment—"I hope I ain't like a miserly fellow we had in our lumber-camp. He was awful pious about some things, and awful mean about others. So the boys said, 'he kept the Sabbath and everything else he could lay his hands upon.' He used to get riled at it.

"Not that I've a word to say against keeping Sunday," went on Herb, in a different key. "Tell you what, out here a fellow thinks a heap of his day o' rest, when his legs can stop tramping, and his mind get a chance to do some tall thinking. Now, boys, we've covered twelve good miles since we left Millinokett Lake, and you needn't go any farther to-day unless you've a mind to. We can make camp right here, near that stream. It will be nice, cold drinking-water, for it has meandered down from Katahdin."

He pointed to a brook a little way ahead, shimmering in the rays of the afternoon sun, of which they caught stray peeps through the gaps in an intervening wall of pines and hemlocks. A few minutes brought them to its brink. Tired and parched from their journey, each one stooped, and quenched his thirst with a delicious, ice-cold draught.

"Was there ever a soda-fountain made that could give a drink to equal that?" said Cyrus, smacking his lips with content. "But listen to the noise this stream makes, boys. I guess if I were to lie beside it for an hour, I'd think, as the Greenlanders do, that I could hear the spirits of the world talking through it."

"That's a mighty queer notion," answered Herb; "and I never knew as other folks had got hold of it. But, sure's you live! I've thought the same thing myself lots o' times, when I've slept by a forest stream. Who'll lend a helping hand in cutting down boughs for our fire and bed? I want to be pretty quick about making camp. Then we'll be able to try some moose-calling after supper."

At this moment a peculiar gulping noise in Neal's throat drew the eyes of his companions upon him. His were bright and strained, peering at the opposite bank.

"Look! What is it?" he gasped, his low voice rattling with excitement.

"A cow-moose, by thunder!" said Herb. "A cow-moose and a calf with her! Here's luck for ye, boys!"

One moment sooner, simultaneously with Neal's gulp of astonishment, there had emerged from the thick woods on the other bank a brown, wild-looking, hornless creature, in size and shape resembling a big mule, followed by a half-grown reproduction of herself.

Her shaggy mane flew erect, her nostrils quivered like those of a race-horse, her eyes were starting with mingled panic and defiance.

A snort, sudden and loud as the report of a shot-gun, made the four jump. Neal, who was standing on a slippery stone by the brink, lost his balance and staggered forward into the water, kicking up jets of shining spray. The snort was followed by a grunt, plaintive, distracted, which sounded oddly familiar, seeing that it had been so well imitated on Herb's horn.

And with that grunt, the moose wheeled about and fled, making the air swish as she cut through it, followed by her young, her mane waving like a pennon.

"Well, if that ain't bang-up luck, I'd like to know what is," said the guide, as he watched the departure. "I never s'posed you'd get a chance to see a cow-moose; she's shyer'n shy. Say! don't you boys think that I've done her grunt pretty well sometimes?"

"That you have," was the general response. "*We* couldn't tell any difference between your noise and the real thing."

"But she wasn't a patch on the bull-moose in appearance," lamented Dol.

"No more she was, boy. Most female forest creatures ain't so good-looking as the males! And that's queer when you think of it, for the girls have the pull over us where beauty is concerned. We ain't in it with 'em, so to speak."

There was a big gale of laughter over Herb Heal's gallant admiration for the other sex, and the sigh which accompanied his expression of it. He joined in the mirth himself, though he walked off to make camp, muttering: —

"Sho! You city fellows think that because I'm a woodsman I never heard of lovemaking in my life."

"Perhaps there is a little girl at some settlement waiting for a home to be fixed up out of guide's fees," retorted Cyrus.

And the three shouted again for no earthly reason, save that the stimulus of forest air and good circulation was driving the blood with fine pressure through their veins, and life seemed such a glorious, unfolding possession — full of a wonderful possible — that they must hold a sort of jubilee.

Herb, who perhaps in his lonely hours in the woods did cherish some vision such as Cyrus suggested, was so infected with their spirit, that, as he swung his axe with a giant's stroke against a hemlock branch, he joined in with an explosive: —

"Hurrup! Hur-r-r-rup!"

This startled the trio like the bursting of a bomb, and trebled their excitement; for their guide, when abroad, had usually the cautious, well-controlled manner of the still-hunter, who never knows what chances may be lurking round him which he would ruin by an outcry.

"Quit laughing, boys," he said, recovering prudence directly he had let out his yell. "Quit laughing, I say, or we may call moose here till crack o' doom without getting an answer. I guess they're all off to the four winds a'ready, scared by our fooling."

## CHAPTER XIX.

#### TREED BY A MOOSE.

"I TOLD you so, boys," breathed the guide two hours later, with an overwhelming sigh of regret, after he had given his most fetching calls in vain. "I told you so. There ain't anything bigger'n a buck-rabbit travelling. That tormented row we made scared every moose within hearing."

Herb was standing on the ground, horn in hand, screened by the great shadows of a clump of hemlocks; the three were perched upon branches high above him, a safe post of observation if any moose had answered.

"You may as well light down now," he continued, turning his face up, though the boys were invisible; "I ain't a-going to try any more music to-night. I guess we'll stretch ourselves for sleep early, to get ready for a good day's work to-morrow. An eight-mile tramp will bring us to the first heavy growth about the foot of Katahdin, and I'll promise you a sight of a moose there."

His companions dropped to earth; and the four sought the shelter of their tent, which had been pitched a few hundred yards from the calling-place. Some dull embers smouldered before it; for Herb, even while preparing supper, had kept the camp-fire very low, lest any wandering clouds of smoke should interfere with the success of his calling.

Now he heaped it high, throwing on without stint withered hemlock boughs and massive logs, which were soon wrapped in a sheet of flame, making an isle of light amid a surrounding sea of impenetrable darkness.

Many times during the night the watchful fellow arose to replenish this fire, so that there might be no decrease in the flood of heat which entered the tent, and kept his charges comfortable. Once, while he was so engaged, the placid sleepers whom he had noiselessly quitted were aroused to terror — sudden, bewildering night-terror — by a gasping cry from his lips, followed by the leaping and rushing of some brute in flight, and by a screech which was one defiant note of unutterable savagery.

"Good heavens! What's that?" said Cyrus.

"Is it — can it — could it be a panther?" stammered Dol.

"Get out!" answered Neal contemptuously. "The panthers have got out long ago, so every one says."

"A lynx! A Canada lynx, boys, as sure as death and taxes!" panted Herb Heal, springing into the tent on the instant, with a burning brand in his hand. "'Tain't any use your tumbling out, for you won't see him. He's away in the thick of the woods now."

Cyrus gurgled inarticulate disappointment. At the first two words he had sprung to his legs, having never encountered a lynx.

"The brute must have been prowling round our tent," went on Herb, his voice thick from excitement. "He leaped past me just as I was stooping to fix the fire, and startled me so that I guess I hollered. He got about half a dozen yards off, then turned and crouched as if he was going to spring back. Luckily, the axe was lying by me, just where I had tossed it down after chopping the last heap of logs. I caught it up, and flung it at him. It struck him on the side, and curled him up. I thought he was badly hurt; but he jumped the next moment, screeched, and made off. A pleasant scream he has; sounds kind o' cheerful at night, don't it?"

No one answered this sarcasm; and Herb flung himself again upon his boughs, pulling his worn blanket round him, determined not to relinquish his night's sleep because a lynx had visited his camp. The city fellows sensibly tried to follow his example; but again and again one of them would shake himself, and rise stealthily, convinced that he heard the blood-curdling screech ringing through the silent night.

It was nearly morning before fatigue at last overmastered every sensation, and the three fell into an unbroken sleep, which lasted until the sun was high in the sky. When they awoke, their sense of smell was the first sense to be tickled. Fragrant odors of boiling coffee were floating into the tent. One after another they scrambled up, threw on their coats, and hurried out to find their guide kneeling by the camp-fire on the very spot from which he had hurled his axe at the lynx a few hours before. But now his right hand held a green stick, on which he was toasting some slices of pork into crisp, appetizing curls.

"'Morning, boys!" he said, as the trio appeared. "Hope your early rising won't opset ye! If you want to dip your faces in the stream, do it quick, for these dodgers are cooked."

The "dodgers" were the familiar flapjacks. Herb set down his stick as he spoke to turn a batch of them, which were steaming on the frying-pan, tossing them high in air as he did so, with a dexterous turn of his wrist.

The boys having performed hasty ablutions in the stream, devoted themselves to their breakfast with a hearty will. There was little leisure for discussing the midnight visit of the lynx, or for anything but the joys of satisfying hunger, and taking in nutrition for the day's tramp, as Herb was in a hurry to break camp, and start on for Katahdin. The morning was very calm; there seemed no chance of a wind springing up, so the evening would probably be a choice one for moose-calling.

In half an hour the band was again on the march, the business of breaking camp being a swift one. The tent was on Herb's shoulders; and naught was left to mark the visit of man to the humming stream but a bed of withering boughs on which the lynx might sleep to-night, and a few dying embers which the guide had thrashed out with his feet.

No halt was made until four o'clock in the afternoon. Then Herb Heal came to a standstill on the edge of a wide bog. It lay between him and what he called the "first heavy growth;" that is, the primeval forest, unthinned by axe of man, which at certain points clothes the foot of Katahdin.

The great mountain, dwelling-place of Pamolah, cradle of the flying Thunder and flashing Lightning, which according to one Indian legend are the swooping sons of the Mountain Spirit, now towered before the travellers, its base only a mile distant.

"I've a good mind to make camp right here," said Herb, surveying the bog and then the firm earth on which he stood. "We may travel a longish ways farther, and not strike such a fair camping-ground, unless we go on up the side of the mountain to that old home-camp I was telling you about, which we built when we were trapping. I guess it's standing yet, and 'twould be a snug shelter; but we'd have a hard pull to reach it this evening. What d'ye say, boys?"

"I vote for pitching the tent right here," answered Cyrus.

The English boys were of the same mind, and the guide forthwith unstrapped his heavy pack-basket. As he hauled forth its contents, and strewed them on the ground, the first article which made its appearance was the moose-horn; it had been carefully stowed in on top. Dol snatched it up as a dog might snatch a bone, and touched it with longing in every finger-tip.

"There's one bad thing about this place," grumbled Herb presently, surveying the landscape wherever his eye could travel, "there isn't a pint of drinking-water to be seen. There may be pools here and there in that bog; but, unless we want to keel over before morning, we'd better let 'em alone. Say! could a couple of you fellows take the camp-kettle, and cruise about a bit in search of a spring?"

"I volunteer for the job!" cried Dol instantly, with the light of some sudden idea shining like a sunburst in his face.

"You don't budge a step, old man, unless I go with you," said Cyrus. "Not much! I don't want to patrol the forests like a lunatic for five mortal hours in search of you, and then find you roasting your shins by some other fellow's camp-fire. One little hide-and-seek game of that kind was enough."

"Well! the fact that I did bring up by Doc's camp-fire shows that I am able to take care of myself. If I get into scrapes, I can wriggle out of them again," maintained the kid of the camp, with a brazen look, while his eyes showed flinty sparks, caused by the inspiring purpose hidden behind them, which had little to do with water-carrying.

"Why can't you both go without any more palaver?" suggested Herb, as he started away towards a belt of young firs to cut stakes for the tent. "Cruise straight across the bog, mark your track by the bushes as you go 'long, don't get into the woods at all, and 'twill be plain sailing. I guess you'll strike a spring before very long."

Cyrus caught up the camp-kettle, and stepped out briskly over the springy, spongy ground. Dol Farrar followed him. The two were half-way across the bog before the elder noticed that the younger was carrying something. It was the moose-horn.

"If we run across any moose-signs, I'm going to try a call," said Dol, his strike-a-light eyes fairly blazing while he disclosed his purpose. "You may laugh, Cy, and call me a greenhorn; but I bet you I'll get an answer, at least if there's a bull-moose within two miles."

"That's pretty cheerful," retorted the Boston man; "especially as neither of us has brought a rifle. Mr. Moose may be at home, and give you an answer; but there's no telling what sort of temper he'll be in."

"I left my Winchester leaning against a tree on the camping-ground," said the would-be caller regretfully. "But you know you wouldn't fire on him, Cy, unless he came near making mince-meat of us. If he should charge, we could make a dash for the nearest trees. Let's risk it if we run across any tracks!"

"And in the meantime, Herb will be wondering where we are, vowing vengeance on us, and waiting for the kettle while we're waiting for the moose," argued Garst. "It won't do, Chick. Give it up until later on. We undertook the job of finding water, and we're bound to finish that business first."

"If I wait until later on, I may wait forever," was the boy's gloomy protest. "To-night, when Herb is there, Neal and you will just sit on me, and be afraid of my making a wrong sound, and spoiling the sport.

"And I *know* we'll see moose-tracks before we get back to camp!" wound up the young pleader passionately. "I've been working up to it all day. I mean I've felt as if something — something fine — was going to happen, which would make a ripping story for the Manchester fellows when we go home. Do let me have one chance, Cy, — one fair and honest chance!"

There was such a tremendous force of desire working through the English boy that it set his blood boiling, and every bit of him in motion. His eyes were afire, his eyelids shut and opened with their quick snap, his lips moved after he had finished speaking, his fingers twitched upon the moose-horn.

He was a picture of heart-eagerness which Cyrus could not resist, though he shook with laughter.

"I'll take mighty good care that the next time I go to find water for the camp-supper, I don't take a crank with me, who has gone mad on moose-calling," he said. "See here! If we do come across moose-signs, I'll get under cover, and give you quarter of an hour to call and listen for an answer — not a second longer. Now stop thinking about this fad, and keep your eyes open for a spring."

But, unfortunately, this seemed to be a thirsty and tantalizing land for travellers. The soft sod under their feet oozed moisture; slimy, stagnant bog-pools appeared, but not a drop of pure, gushing water, to which a parched man dare touch his lips.

They crossed the wide extent of bog, Cyrus breaking off stunted bushes here and there to mark his pilgrimage; they reached the dense timber-growth at the base of the mountain, longing for the sight of a spring as eagerly as ever pilgrims yearned to behold a healing well; but their search was unsuccessful.

Decidedly nonplussed, Dol all the time keeping one eye on the lookout for water and the other for moose-signs, they took counsel together, and determined to "cruise" to the right, skirting the foot of Katahdin, hoping to find a gurgling, rumbling mountain-torrent splashing down. Having travelled about half a mile in this new direction, with the giant woods which they dared not enter rising like an emerald wall on the one hand, and the dreary bog-land on the other, they at last, when patience was failing, came to a change in the landscape.

The desired water was not in view yet; but

the bog gave way to fairer, firmer ground, covered with waving grasses, studded with rising knolls, and having no timber growth, save stray clumps of birches and hemlocks, several hundred yards apart.

"Now, this is jolly!" exclaimed Dol. "This looks a little bit like an English lawn, only I'm afraid it's not a likely place for moose-tracks. But I'm glad to be out of that beastly bog."

"Confusion to your moose-tracks," ejaculated Cyrus, half exasperated. "I wish we could find a well. That would be more to the purpose. Listen, Dol, do you hear anything?"

"I hear — I hear — 'pon my word! I *do* hear the bubbling and tinkling of water somewhere! Where on earth is it? Oh! I know. It comes from that knoll over there — the one with the bushes."

Dol Farrar, as he finished his jerky sentences, pointed to an eminence which was two or three hundred yards from where they stood, and a like distance from the wall of forest.

"Well! It's about time we struck something at last," grumbled Garst. "Catch me ever coming on a water pilgrimage again!

I'll let Herb fill his own kettle in future. Now, I believe that fellow could smell a spring."

"Just as I smelt this one!" exclaimed Dol triumphantly. "I told you 'twas on the side of the knoll. And here it is!"

"Bravo, Chick! You've got good ears, if you are crazy upon one subject."

And so speaking, Cyrus, with a chuckle of joy, unslung the tin drinking-cup which hung at his belt, filled and refilled it, drinking long, inspiring draughts before he prepared to fill the camp-kettle.

"The best water I ever tasted, Dol!" he exclaimed, smacking his lips. "It's ice-cold. There's not much of it, but it has quality, if not quantity."

The long-sought well was, in truth, a tiny one. It came bubbling up, clear and pellucid, from the bowels of the earth, and showed its laughing face amid a cluster of bushes — which all bent close to look at it lovingly — half-way up the knoll. A wee stream trickled down from it, — dribble — dribble — a rivulet that had once been twice its present size, judging from the wide margin of spattered clay at each side.

Dol had been following his companion's

example, and drinking joyfully before thinking of aught else. When the moment came for him to straighten his back, and rise upon his legs, instead of this natural proceeding, he suddenly crouched close to the ground, his breath coming in quick puffs, his eyes dilating, a froth of excitement on his lips.

"What on earth are you staring at?" asked Cyrus. "You look positively crazy."

For answer, the English boy shot up from his lowly posture, seized his companion by the arm, making him drop the camp-kettle, which he was just filling, and forced him to scan the soft clay by the rivulet.

"Look there — and there!" gurgled Dol, his voice sounding as if he was being choked by suppressed hilarity. "I told you we'd find them, and you didn't believe me! Aren't those moose-tracks? They're not deer-tracks, anyhow; they're too big. I may be a greenhorn, but I know that much."

"They *are* moose-tracks," Cyrus answered slowly, almost unbelievingly, though the evidence was before him. "They certainly are moose-tracks," he repeated, "and very recent ones too. A moose has been drinking here, perhaps not half an hour ago. He can't be far away."

Garst was now warming into excitement himself. His bass tones became guttural and almost inarticulate, while he lowered them to prevent their travelling. On the reddish clay at his feet were foot-marks very like the prints of a large mastiff. He studied them one by one, even tracing the outline with his forefinger.

"Then I'm going to call," whispered Dol, his words tremulous and stifled. "Lie low, Cy! You promised you'd give me a fair chance; you'll have to keep your word."

"I'll do it too," was the answering whisper. "But let's get higher up on the knoll, behind those big bushes at the top. And listen, Dol, if a moose makes a noise anywhere near, we must scoot for the trees before he comes out from cover. I've got to answer to your father for you."

It was an intense moment in Dol Farrar's life; sensation reached its highest pitch, as he crouched low behind a prickly screen, put the birch-bark horn to his mouth, and slowly breathed through it with the full power of his young lungs, marvellously strengthened by the forest life of past weeks.

There was a minute's interval while he removed it again, and drew in all the air he could contain. Then a call rose upon the evening air, so touching, so plaintive, with such a rising, quavering impatience as it surged out towards the woods, — whither the boy-caller's face was turned, — that Cyrus could scarcely suppress a " Bravo!"

The summons died away in a piteous grunt. A second time the call rose and fell. On the third repetition it broke off, as usual, in an abrupt roar, which seemed to strike the tops of the giant trees, and boom among them.

A froth was on Dol Farrar's lips, his eyes were reddened, he puffed hard through spread nostrils, like a young horse which has been trying its mettle for the first time, as he lowered that moose-horn, lifted his head, and cocked his ears to listen.

Two soundless minutes passed. Dol, who, if he had mastered the hunter's call, had certainly not mastered his patience, put the bark-trumpet again to his lips, determined to try the effect of a surpassingly expressive grunt.

But he never executed this false movement, which would have given away the trick at once.

A bellow — a short, snorting, challenging bellow — burst the silence, coming from the very edge of the woods. It brought Cyrus to his feet with a jump. It so startled the ambitious moose-caller, that, in rising hurriedly from his squatting position, he lost his balance, and rolled over and over to the bottom of the knoll, smashing the horn into a hundred pieces.

He picked himself up unhurt, but with a sensation as if all the bells in Christendom were doing a jumbled ringing in his head. And loud above this inward din he heard the sound, so well remembered, as of an axe striking repeatedly against a tree, the terrible chopping noises of a bull-moose, not two hundred yards away.

No sooner had he scrambled to his legs, than Garst was at his side, gripping his arm, and forcing him forward at a headlong run.

"You've done it this time with a vengeance!" bawled the Bostonian. "He's coming for us straight! And we without our rifles! The trees! The trees! It's our only chance!"

With the belling still in his head, and so bewildered by his terrible success that he felt as if his senses were shooting off hither and thither like rockets, leaving him mad, Dol nevertheless ran as he had never run before, shoulder to shoulder with his comrade, dashing wildly for a clump of hemlocks over a hundred yards distant. Yet, for the life of him, he could not help glancing back once over his shoulder, to see the creature which he had humbugged, luring it from its forest shelter, and which now pursued him.

The moose was charging after them full tilt, gaining rapidly too, his long thin legs, enormous antlers, broad, upreared nose, and the green glare in his starting eyes, making him look like some strange animal of a former earth. Dol at last trembled with actual fear. He gave a shuddering leap, and forced his legs, which seemed threatened with paralysis, to wilder speed.

"Climb up that hemlock! Get as high as you can!" shrieked Cyrus, stopping to give him an upward shove as they reached the first friendly trunk.

Dol obeyed. Gasping and wild-eyed, he dug his nails into the bark, clambering up somehow until he reached a forked branch about eight feet from the ground. Here strength failed. He could only cling dizzily, feeling that he hung between life and death.

The moose was now snorting like a war-horse beneath. The brute stood off for a minute, then charged the hemlock furiously, and butted it with his antlers till it shook to its roots, the sharp prongs of those terrible horns coming within half an inch of Dol's feet.

With a gurgle of horror the boy tried to reach a higher limb, and succeeded; for at the same moment a timely shout encouraged him. Cyrus was bawling at the top of his voice from a tree ten feet distant: —

"Are you all right, Dol? Don't be scared. Hold on like grim death, and we can laugh at the old termagant now."

"I'm — I'm all right," sang out Dol, though his voice shook, as did every twig of his hemlock, which the moose was assaulting again. "But he's frantic to get at me."

"Never mind. He can't do it, you know. Only don't you go turning dizzy or losing your balance. Ha! you old spindle-legged monster, stand off from that tree. Take a turn at mine now, for a change. You can't shake me down, if you butt till midnight."

Garst's last sentences were hurled at the moose. The Bostonian, having reached a safe height, thrust his face out from his screen of branches, waving first an arm, and then a leg, at the besieging foe, hoping that the force of those battering antlers would be directed against his hemlock, so that his friend's nerves might get a chance to recover.

The ruse succeeded. The moose, reminded that there was a second enemy, charged the other tree; stood off for a minute to get breath, then charged it again, snorting, bellowing, and knocking his jaws together with a crunching, chopping noise.

"Ha! that's how he makes the row like a man with an axe — by hammering his jaws on each other. Well, well! but this is a regular picnic, Dol," sang out Cyrus jubilantly, caring nothing for the shocks, and forgetting camp, water, peril, everything, in his joy at getting a chance to leisurely study the creature he had come so far to visit.

"I owe you something for this, little man!" he carolled on in triumph, as he watched every wild movement of the moose. "This is a show we'll only see once in our lives. It's worth a hundred dollars a performance. Butt and snort till you're tired, you 'Awful Jabberwock!'"— this to the bull-moose. "We've come hundreds of miles to see you, and the more you carry on the better we'll be pleased."

Indeed, the wrathful king of forests seemed in no hurry to cut short his pantomime. He ramped and raged, tearing from one tree to another, expending paroxysms of force in vain attempts to overturn one or the other of them. The ground seemed to shake under his thundering hoofs. His eyes were full of green fire; his nostrils twitched; the black tassel or "bell" hanging from his shaggy throat shook with every angry movement; his muffle, the big overhanging upper lip, was spotted with foam.

As he gulped, grunted, snorted, and roared, his uncouth, guttural noises made him seem more than ever like a curious creature of earth's earliest ages.

"We came pretty near to being goners, Dol, I tell you!" carolled Cyrus again from his high perch in the hemlock, carrying on a by-play with the enemy between each sentence. "How in the name of wonder did you manage such a call? It would have moved the heart-strings of any moose. I was lying flat, you know, peeping through a little gap in the bushes, and you had scarcely taken the horn from your mouth when I saw the old fellow come stamping out of the woods. My! wasn't he a sight? He stood for a minute looking about for the fancied cow; then he bellowed, and started towards the knoll. I knew we had better run for our lives. As soon as he saw us he gave chase."

"And 'the fancied cow' should go tumbling down the knoll like a rolling jackass, and smash that grand horn to bits!" lamented Dol, who now sat serenely on his bough, with a firm clasp of the hemlock trunk, and a reckless enjoyment of the situation which far surpassed his companion's.

Cyrus began to have an occasional twinge of uneasiness about the possible length of the siege, after his first exuberance subsided; but the younger boy, his short terror overcome, had no misgivings. He coquetted with the moose through a thick screen of foliage, shook the branches at him, gibed and taunted him, enjoying the extra fury he aroused.

But suddenly the old bull, having kept up his wild movements for nearly an hour, resolved on a change of tactics. He stood stock-still and lowered his head.

"Goodness! He has made up his mind to 'stick us out!'" gasped Cyrus.

"What's that?" said Dol.

"Don't you see? He's going to lay siege in good earnest — wait till we're forced to come down. Here's a state of things! We can't roost in these trees all night."

The hemlocks were throwing ever-lengthening shadows on the grass. A slow eclipse was stealing over everything. The motionless moose became an uncouth black shape. Garst muttered uneasily. His fingers tingled for his rifle — a very unusual thing with him. His eyes peered through the creeping darkness in puzzled search for some suggestion, some possibility of escape.

"If it were only myself!" he whispered, as if talking to his hemlock. "If it were only myself, I wouldn't care a pin. 'Twould do me no great harm to perch here for hours. But an English youngster, on his first camping-trip! Why, the chill of a forest night might ruin him. He wouldn't howl or make a fuss, for both those Farrar boys have lots of grit, but he'd never get over it. Dol!" he wound up, raising his voice to a sharp pitch. "Say, Dol, I'm going to try a shout for help. Herb must be getting anxious about us by this time. If we could once make him hear, he could try some trick to lure this old curmudgeon away, or creep up and shoot him. Something must be done."

Fetching a deep breath, Cyrus sent a distance-piercing "Coo-hoo!" ringing through the night-air. He followed it with another.

But, so far as he could hear, the hails fetched no answer, save from the moose-jailer. The brute was stirred into a fresh tantrum by the noise. He charged the hemlocks once more, butted and shook them like a veritable demon.

When his paroxysm had subsided, and he stood off to get breath, Garst hailed again.

Glad sound! An answer this time! First, a shrill, long "Coo-hoo!" Next, Herb's voice was heard pealing from far away in the bog: "What's up, boys? Where in the world are you?"

"Here in the trees — treed by a bull-moose!" yelled Cyrus. "He's the maddest old monster you ever saw. Could you coax him off, or sneak up and shoot him? He means to keep us prisoners all night."

There was no wordy answer. But presently the treed heroes heard an odd, bird-like whistle. Dol thought it came from a feathered creature; his more experienced companion guessed that the guide's lips gave it as a signal that he was coming, but that he didn't want to draw the moose's attention in his direction just yet.

Such a quarter of an hour followed! With the fresh spurt of anger the bull-moose became more savage than ever. He grunted, tramped, and hooked the trees with his horns, so that the pair who were perched like night-birds on the branches had to hold on for dear life, lest a surprising shock should dislodge them. Whenever the creature stood off, to gather more fury, they could have counted their heart-beats while they listened, breathlessly anxious to know what action the approaching woodsman would take.

Once Cyrus spoke.

"Dol Farrar," he said, "I guess this caps all the adventures that you or I have had up to date. No wonder you felt all day as if you were working up to something. I'll believe in presentiments in future."

The words had scarcely passed his lips, when there was the sharp bang! bang! of a rifle not twenty yards distant. A bright sputter of fire cut the darkness beneath the hemlocks.

The moose's blind rage threatened to be his own undoing. While he was fighting an imaginary danger, ears and nostrils half-choked by fury, through the calm night Herb Heal, Winchester in hand, had crept noiselessly on, till he reached the very trees which sheltered his friends.

Once, twice, three times the rifle snapped. The first shot missed altogether. At the second, the moose rose upon his hind-legs, with a sharp sound of fright and pain, quite unlike his former noises. Then he gave a quick jump.

"Great Governor's Ghost! he's gone;" yelled Cyrus, who had swung himself down a few feet, and was hanging by one arm, in his anxiety to see the result of the firing. "You needn't shoot again, Herb! He's off! Let him go!"

"I guess that second shot cut some hair from him, and drew blood too," answered Herb, his deep voice giving the pair a queer sensation as they heard it right beneath. "It was too dark to see plain, but I think he reared; and that's a sign that he was hurt, little or much. Don't drop down for a minute, boys, till we see whether he has bolted for good."

## CHAPTER XX.

#### DOL.'S TRIUMPH.

HE had bolted for good, vanished into the mysterious deeps of the primeval forest, whether hurt unto death, or merely "nipped" in a fore-leg, as Herb inclined to think, nobody knew.

"It's too dark to see blood-marks, if there are any, so we can't trail him to-night. If he's hit bad — but I guess he ain't — we can track him in the morning," said the guide; as, after an interval of listening, the rescued pair dropped down from their perches. "Did he chase you, boys? Where on earth did you come on him?"

Talking together, their words tumbling out like a torrent let loose, Cyrus Garst and Dol

Farrar gave an account of the past two hours — strangest hours of their lives — filling up the picture of them bit by bit.

"Whew! whew! You did have a narrow squeak, boys, and a scarey time; but I guess you had a lot of fun out of the old snorter," said Herb, his rare laugh jingling out, starting the forest echoes like a clang of bells. "You've won those antlers, Dol — won 'em like a man. Blest, but you have! I promised 'em to the first fellow who called up a moose; and nary a woodsman in Maine could have done it better. I'm powerful glad 'twasn't your own death-call you gave. I'll keep my eye on you now till you leave these woods. Where's the horn?"

"Smashed to bits," answered Dol regretfully.

"And the camp-kettle?"

"Lying by the spring, over there on the knoll, unless the moose kicked it to pieces," said Cyrus.

"My senses! you're a healthy pair to send for water, ain't ye? Let's cruise off and find it. I guess you'll be wanting a drink of hot coffee, after roosting in them trees for so long."

Garst led the way to the spring. Its

pretty hum sounded like an angel's whisper through the night, after the tumult of the past scene. Herb fumbled in his leather wallet, brought out a match and a small piece of birch-bark, and kindled a light. With some groping, the kettle was found; it was filled, and the party started for camp.

"I heard the distant challenge of a bull-moose a couple of hours ago," said the guide, as they went along. "I never suspicioned he was attacking you; but after the camp was a' ready, and you hadn't turned up, I got kind o' scared. I left Neal to tend the fire and toast the pork, and started out to search. I s'pose I took the wrong direction; for I hollered, and got no answer. Afterwards, when I was travelling about the bog, I heard a 'Coo-hoo!' and the noises of an angry moose. Then I guessed there was trouble."

"Won't Neal look blue when he hears that he was toasting pork while we were perched in those trees, with the moose waltzing below!" exclaimed Dol. "Well, Cy, I've won the antlers, and I've got my ripping story for the Manchester fellows. I don't care how soon we turn home now."

"You don't, don't ye?" said the guide. "Well, I should s'pose you'd want to trail up that moose to-morrow, and see what has become of him."

"Of course I do! I forgot that."

And Dol Farrar, who had thought his record of adventure and triumph so full that it could hold no more, realized that there is always for ambition a farther point.

Neal did feel a little blue over the thought of what he had missed. But, being a generous-hearted fellow, he tasted his young brother's joy, when the latter cuddled close to him upon the evergreen boughs that night, muttering, as if the whole earth lay conquered at his feet: —

"My legs are as stiff as ramrods, but who'd think of his legs after such a night as we've had?

"I say, Neal, this is life; the little humbugging scrapes we used to call adventures at home are only play for girls. It's something to talk about for a lifetime, when a fellow comes to close quarters with a creature like that moose. I said I'd get the better of his ears, and I did it. Pinch me, old boy, if I begin a moose-call in my sleep."

Several times during the night Neal found it necessary to obey this injunction, else had there been no peace in the camp. But, in spite of Dol's ravings and riotings in his excited dreams, the party enjoyed a needed ten hours' slumber, all save Herb, who, as usual, was astir the next morning while his comrades were yet snoring.

He got his fire going well, and baked a great flat loaf of bread in his frying-pan, setting the pan amid hot ashes and covering it over. Previous to this, he had made a pilgrimage to the distant spring, to fill his kettle for coffee and bread-making, and had carefully examined the ground about the clump of hemlocks.

The result of his investigation was given to the boys as they ate their breakfast under the shade of a cedar, with a sky above them whose morning glories were here and there overshot by leaden tints.

"I guess we've got a pretty fair chance of trailing that moose," he said. "I found both hair and blood on the spot where he was wounded. I'm for following up his tracks, though I guess they'll take us a bit up the mountain. If he's hurt bad, 'twould be kind o' merciful to end his sufferings. If he ain't, we can let him get off."

"Right, as you always are, Herb," answered Cyrus. "But what on earth made the creature bolt so suddenly? If you had seen him five minutes before he was shot, you'd have said he had as much fight in him as a lion."

"That's the way with moose a'most always. Their courage ain't that o' flesh-eating animals. It's only a spurt; though it's a pretty big spurt sometimes, as you boys know now. It'll fail 'em in a minute, when you least expect it. And, you see, that one last night didn't know where his wound came from. I guess he thought he was struck by lightning or a thunder-ball, so he skipped. Talking of thunder-balls, boys," wound up Herb, "I shouldn't be surprised if the old Mountain Spirit, who lives up a-top there, gave us a rattling welcome with his thunders to-day. The air is awful heavy for this time of year. Perhaps we'd better give up the trailing after all."

"Nonsense!" exclaimed Dol indignantly. "Do you think a shower will melt us? Or that we'll squeal like girls at a few flashes of lightning? 'Twould be jolly good fun to see old Pamolah sending off his artillery."

"Well, there'd be no special danger, I guess, if we were past the heavy timber growth before the storm began. There's lots of rocky dens on the mountain side where we could shelter under a granite ledge, and be safer than we'd be here in tent. Or we might come a-near our old log camp. I guess, if that's standing yet, you'd like to see it. Say! we'll leave it to Cyrus. He's boss, ain't he?"

Cyrus, desperately anxious to know whether it would be life or death for the wounded moose, and regarding the signs of bad weather as by no means certain, decided in favor of the expedition. The campers hurriedly swallowed the remainder of their breakfast, and made ready for an immediate start.

"In trailing a moose the first rule is: go as light as you can; that is, don't carry an ounce more stuff than is necessary. Even a man's rifle is apt to get in his way when he has to scramble over windfalls, or slump between big bowlders of rock, which a'most tear the clothes off his back. And we may have to do some pretty tall climbing. So leave all your traps in the tent, boys; I'll fasten it down tight. There won't be any human robbers prowling around, you bet! Bears and coons are the only burglars of these woods, and they don't do much mischief in daytime."

The guide rapidly gave these directions,

his breezy voice setting a current of energy astir, like a wind-gust cutting through a quiet grove, while he rolled his indispensable axe, some bread that was left from the meal, and a lump of pork into a little bundle, which he strapped on his back.

"Now," he said, "if that trail should give us a long tramp, or if you boys should take a notion to go a good ways up Katahdin, or anything turns up to hinder our getting back to camp till nightfall, I've our snack right here. I can light a fire in two minutes, to toast our pork; and we'll wash it down with mountain water, the best drink for climbers. I could rig you up a snug shelter, too, in case of accidents. A woodsman ain't in it without his axe."

To what strange work that axe would be put ere night again closed its shutters over granite peaks and evergreen forest, Herb Heal little knew; nor could he have guessed that the coming hours would make the most heart-stirring day of his stirring life. If he could, would he have started out this morning with a happy-go-lucky whistle, softly modulated on his lips, and no more sober burden on his mind than the trail of that moose?

## CHAPTER XXI.

#### ON KATAHDIN.

"SEE there, boys, I told you so," said Herb, as the party reached the ever-to-be-remembered clump of hemlocks, the beginning of the trail which they were ready to follow up like sleuth-hounds. "There's plenty of hair; I guess I singed him in two places."

He pointed to some shaggy clotted locks on the grass at his feet, and then to a small maroon-colored stain beside them.

"Is that blood?" asked Neal.

"Blood, sure enough, though there ain't much of it. But I'll tell you what! I'd as soon there wasn't any. I wish it had been light enough last night for me to act barber, and only cut some hair from that moose, instead of wounding him. It might have answered the purpose as well, and sent him walking."

"I don't believe it would have done anything of the kind," exclaimed Dol. "He was far too red-hot an old customer to bolt because a bullet shaved him."

"Well, I don't set up to be soft-hearted like Cyrus here; and I'm ready enough to bag my meat when I want it," said the woodsman. "But sure's you live, boys, I never wounded a free game creature yet, and seed it get away to pull a hurt limb and a cruel pain with it through the woods, that I could feel chipper afterwards. It's only your delicate city fellows who come out here for a shot once a year, who can chuckle over the pools of blood a wounded moose leaves behind him. Sho! it's not manly."

A start was now made on the trail, Herb leading, and showing such wonderful skill as a trailer that the English boys began to believe his long residence in the woods had developed in him supernatural senses.

"That moose was shot through the right fore-leg," he whispered, as the trackers reached the edge of the forest.

"How do you know?" gasped the Farrars.

The woodsman answered by kneeling, bending his face close to the ground, and drawing his brown finger successively round three prints on a soft patch of earth, which the unpractised eyes could scarcely discern.

"There's no mark of the right fore-hoof," he whispered again presently; "nothing but *that*," pointing to another dark red blotch, which the boys would have mistaken for maroon-tinted moss.

A breathless, wordless, toiling hour followed. Through the dense woods, which sloped steadily upward, clothing Katahdin's highlands, Herb Heal travelled on, now and again halting when the trail, because of freshly fallen pine-needles or leaves, became quite invisible. Again he would crouch close to the ground, make a circle with his finger round the last visible print, and work out from that, trying various directions, until he knew that he was again on the track which the limping moose had travelled before him.

His comrades followed in single file, carrying their rifles in front of their bodies instead of on their shoulders, so that there might be no danger of a sudden clang or rattle from the barrels striking the trees. Following the example of their guide, each one carefully avoided stepping on crackling twigs or dry branches, or rustling against bushes or boughs. The latter they would take gingerly in their hands as they approached them, bend them out of the way, and gently release them as they passed. Heroically they forebore to growl when their legs were scraped by jagged bowlders or prickly shrubs, giving thanks inwardly to the manufacturers of their stout tweeds that their clothes held together, instead of hanging on them like streamers on a rag-bush.

It was a good, practical lesson in moose-trailing; but, save for the knowledge gained by the three who had never stalked a moose before, it was a failure.

The air beneath the dense foliage grew depressing — suffocating. Each one longed breathlessly for the minute when he should emerge from this heavy timber-growth, even to do more rugged climbing. Distant rumbles were heard. Herb's prophecy was being fulfilled. Pamolah was grumbling at the trailers, and sending out his Thunder Sons to bid them back.

But it was too late for retreat. If they gave up their purpose, turned and fled to camp, the storm, which was surely coming, would catch them under the interlacing trees, a danger which the guide was especially anxious to avoid. He pressed on with quickened steps, stooping no more to make circles round the moose's prints. Old Pamolah's threatenings grew increasingly sullen. At last the desired break in the woods was reached; the trackers found themselves on the open side of Katahdin, surrounded by a tangled growth of alders and white birches struggling up between granite rocks; then the mountain artillery broke forth with terrifying clatter.

A loud, long thunder-roll was echoed from crag, slide, forest, spur, and basin. The "home of storms" was a fort of noise.

"Ha! there'll be a big cannonading this time, I guess. Pamolah is going to let fly at us with big shot, little shot, fire and water— all the forces the old scoundrel has," said Herb Heal, at last breaking the silence which had been kept on the trail, and looking aloft towards the five peaks guarding that mysterious basin, from which heavy, lurid clouds drifted down.

At the same time a blustering, mighty wind-gust half swept the four climbers from their feet. A great flash of globe lightning cut the air like a dazzling fire-ball.

"We'll have to quit our trailing, and scoot for shelter, I'm thinking!" exclaimed Cyrus.

"Good land, I should say so!" agreed the guide. "The bull-moose likes thunder. He's away in some thick hole in the forest now, recovering himself. We couldn't have come up with him anyhow, boys, for them blood-spots had stopped. I guess his leg wasn't smashed; and he'll soon be as big a bully as ever. Follow me now, quick! Mind yer steps, though! Them bushes are awful catchy!"

Undazzled by the lightning's frequent flare, unstaggered by the down-rushing wind, as if the mountain thunders were only the roll of an organ about his ears, Herb Heal sprang onward and upward, tugging his comrades one by one up many a precipitous ledge, and pulling them to their feet again when the tripping bushes brought their noses to the ground and their heels into the air.

"Hitch on to me, Dol!" he cried, suddenly turning on that youngster, who was trying to get his second breath. "Tie on to me tight. I'll tow you up! I wish we could ha' reached that old log camp, boys. 'Twould be a stunning shelter, for it has a wall of rock to the back. But it's higher up, and off to the right. There! I see the den I'm aiming for."

A few energetic bounds brought Herb, with Dol in tow, to a platform of rock, which rose above a bed of blueberry bushes. It narrowed into a sort of cave, roofed by an overhanging bowlder.

"We'll be snug enough under this rock!" he exclaimed, pointing to the canopy. "Creep in, boys. We'll have tubs of rain, and a pelting of hail. The rumpus is only beginning."

So it was. The storm had been creeping from its cradle. Now it swept down with an awful whirl and commingling of elements.

The boys, peering out from their rocky nest, saw a magnificent panorama beneath them. The regiments of the air were at war. Lightning chains encircled the heavens, lighting up the forests below. Winds charged down the mountain-side, sweeping stones and bushes before them. Hail-bullets rattled in volleys. Thunder-artillery boomed until the very rocks seemed to shake.

"It's fine!" exclaimed Cyrus. "It's superfine!"

Then a curtain of thick rain partly hid the warfare, the lightning still rioting through it like a beacon of battle.

"The stones up above will have to be pretty firmly fixed to keep their places," said Herb. "Boys, I hope there ain't a-going to be slides on the mountain after this."

"Slides?" echoed Dol questioningly.

"Landslides, kid. Say! if you want to be scared until your bones feel limp, you've got to hear a great big block of granite come ploughing down from the top 'o the mountain, bringing earth and bushes along with it and smashing even the rocks to splinters as it pounds along."

"I guess that's a sensation we'd rather be spared," said Cyrus gravely.

And under the quieting spell of the airy warfare there was silence for a while.

"Do you think it's lightening up, Herb?" asked Neal, after the storm had raged for three-quarters of an hour.

"I guess it is. The rain is stopping too. But we'll have an awful slushy time of it getting back to camp. To plough through them soaked forests below would be enough to give you city fellows a shaking ague."

"Couldn't we climb on to your old log camp?" suggested Garst. "If we have the luck to find the old shanty holding together, we can light a fire there after things dry out a bit, and eat our snack. Then we needn't be in a hurry to get down. We'll risk it, anyhow."

"I reckon that's about the only thing to be done," assented the guide.

And in twenty minutes' time the four were again straining up Katahdin, clutching slippery rocks, sinking in sodden earth, shivering as they were besprinkled by every bush and dwarfed tree, and dreadfully hampered with their rifles.

"Never mind, boys; we'll get there! Clinch yer teeth, and don't squirm! Once we're past this tangle, the bit of climbing that's left will be as easy as rolling off a log!"

So shouted Herb cheerfully, as he tore a way with hand and foot through the stunted growth of alders and birch, which, beaten down by the winds, was now an almost impassable, sopping tangle.

"Keep in my tracks!" he bellowed again. "Gracious! but this sort o' work is as slow as molasses crawling up-hill in winter."

But ten minutes later, when the dripping jungle was behind, he dropped his jesting tone.

He came to a full stop, catching his breath with a big gulp.

"Boys," he cried, "it's standing yet! I see it — the old home-camp! There it is above us on that bit of a platform, with the big rock behind it. And I've kep' saying to myself for the last quarter of an hour that we wouldn't find it — that we'd find nary a thing but mildewed logs!"

A wealth of memories was in the woodsman's eyes as he gazed up at the timber nest, the log camp which his own hands had put up, standing on a narrow plateau, and built against a protecting wall of rock that rose in jagged might to a height of thirty or forty feet.

An earth bank or ridge, covered with hardy mosses and mountain creepers, sloped gently up to the sheltered platform. To climb this was, indeed, " as easy as rolling off a log."

"We used to have a good beaten path here, but I guess it's all growed over," said Herb in a thick voice, as if certain cords in his throat were swelling. "Many's the time I've blessed the sight of that old home-camp, boys, after a hard week's trapping. Hundert's o' night's I've slept snug inside them log walls when blasts was a-sweeping and bellowing around, like as if they'd rip the mountain open, and tear its very rocks out."

While the guide spoke he was leaping up the ridge. A few minutes, and he stood, a towering figure, on the platform above, waving his battered hat in salute to the old camp.

"I guess some traveller has been sheltering here lately!" he cried to Neal Farrar, as the latter overtook him. "There's a litter around," pointing to dry sticks and withered bushes strewn upon the camping-ground. "And the door's standing open. I wonder who found the old shanty?"

Neal remembered, hours afterwards, that at the moment he felt an odd awakening stir in him, a stir which, shooting from head to foot, seemed to warn him that he was nearing a sensation, the biggest sensation of this wilderness trip.

He heard the voices of Cyrus and Dol hallooing behind; but they sounded away back and indistinct, for his ears were bent towards the deserted camp, listening with breathless expectation for something, he didn't know what.

One minute the vague suspense lasted, while he followed Herb towards the hut. Then heaven and earth and his own heart seemed to stand still.

Through the wide-open door of the shanty

came random, crooning snatches of sound. Was the guttural voice which made them human? The English boy scarcely knew. But as the noise swelled, like the moaning of a dry wind among trees, he began, as it were, to disentangle it. Words shaped themselves, Indian words which he had heard before on the guide's tongue.

> "*N'loan pes-saus, mok glint ont-aven,
> Glint ont-aven, nosh morgun.*"

These lines from the "Star Song," the song which Herb had learned from his traitor chum, floated out to him upon Katahdin's breeze. They struck young Farrar's ears in staggering tones, like a knell, the sadness of which he could not at the moment understand. But he had a vague impression that the mysterious singer in the deserted camp attached no meaning to what he chanted.

"Look out, I say! I don't want to come a cropper here."

It was Dol's young voice which rang out shrilly among the mountain echoes. Side by side with Cyrus, the boy had just gained the top of the ridge when the guide suddenly backed upon him, Herb's great shoulder-blade knocking him in the face, so that he had to plant his feet firmly to avoid spinning back.

But Herb had heard that guttural crooning. Just now he could hear nothing else.

Twice he made a heaving effort to speak, and the voice cracked in his throat.

Then, as he sprang for the camp-door, four words stumbled from his lips : —

" By thunder! it's Chris."

## CHAPTER XXII.

#### THE OLD HOME–CAMP.

THE silence which followed that ejaculation was like the hush of earth before a thunder-storm.

Not a syllable passed the lips of the boys as they followed Herb into the log hut, but feeling seemed wagging a startled tongue in each finger-tip which convulsively pressed the rifles.

And not another articulate sentence came from the guide; only his throat swelled with a deep, amazed gurgle as he reached the interior of the shanty, and dropped his eyes upon the individual who raised that queer chanting.

On a bed of withered spruce boughs,

strewn higgledy-piggledy upon the camp-floor — mother earth — lay the form of a man. Thin wisps of blue-black hair, long untrimmed, trailed over his face and neck, which looked as if they were carved out of yellow bone. His figure was skeleton-like. His lips — the lips which at the entrance of the strangers never ceased their wild crooning — were swollen and fever-scorched. His black eyes, disfigured by a hideous squint, rolled with the sick fancies of delirium.

Cyrus and the Farrars, while they looked upon him, felt that, even if they had never heard Herb's exclamation, they would have had no difficulty in identifying the creature, remembering that story which had thrilled them by the camp-fire at Millinokett. It was Herb Heal's traitor chum — the half-breed, Cross-eyed Chris.

And Herb, backing off from the withered couch as far as the limited space of the cabin would allow, stood with his shoulders against the mouldy logs of the wall, his eyes like peep-holes to a volcano, gulping and gurgling, while he swallowed back a fire of amazed excitement and defeated anger, for which his backwoods vocabulary was too cheap.

A flame seemed scorching and hissing

about his heart while he remembered that during some hour of every day for five years, since last he had seen the "hound" who robbed him, he had sworn that, if ever he caught the thief, he would pounce upon him with a woodsman's vengeance.

"I couldn't touch him now — the scum! But I'll be switched if I'll do a thing to help him!" he hissed, the flame leaping to his lips.

Yet he had a strange sensation, as if that vow was broken like an egg-shell even while he made it. He knew that "the two creatures which had fought inside of him, tooth and claw," about the fate of his enemy, were pinching his heart by turns in a last hot conflict.

His eyes shot flinty sparks; he drew his breath in hard puffs; his knotted throat twitched and swelled, while they (the man and the brute) strove within him; and all the time he stood staring in grisly silence at the half-breed.

The latter still continued his Indian croon; though from the crazy roll of his malformed eyes it was plain that he knew not whether he chanted about the stars, his old friends and guides, or about anything else in heaven or earth.

But one thing quickly became clear to Cyrus, and then to the Farrar boys,— less accustomed to tragedy than their comrade,— that this strange personage, in whose veins the blood of white men and red men met, carrying in its turbid flow the weaknesses of two races, was singing his swan-song, the last chant he would ever raise on earth.

At their first entrance, as their bodies interfered with the broad light streaming through the cabin-door, Chris had lifted towards them a scared, shrinking stare. But, apparently, he took them for the shadows which walked in the dreams of his delirium. Not a ray of recognition lightened the blankness of that stare as Herb's big figure passed before him. Letting his eyes wander aimlessly again from log wall to log wall, from withered bed to mouldy rafters, his lips continued their crooning, which sank with his weakening breath, then rose again to sink once more, like the last wind-gusts when the storm is over.

Suddenly his shrunken body shivered in every limb. The humming ceased. His yellow teeth tapped upon each other in trouble and fear. He raised himself to a squatting posture, with his knee-bones to his chin, the wisps of hair tumbling upon his naked chest.

"It's dark — heap dark!" he whimpered, between long gasps. "Can't strike the trail — can't find the home-camp. Herb — Herb Heal — ole pard — 'twas I took 'em — the skins. 'Twas — a dog's trick. Take it out — o' my hide — if yer wants to — yah! Heap sick!"

Not a ray of sense was yet in the half-breed's eyes. An imaginary, vengeance-dealing Herb was before him; but he never turned a glance towards the real, and now forgiving, old chum, who leaned against the wall not ten feet away. His voice dropped to a guttural rumble, in which Indian sounds mingled with English.

But the flame at Herb's heart was quenched at the first whimpered word. His stiffened muscles and lips relaxed. With a gurgle of sorrow, he crossed the camp-floor, and dropped into a crawling position on the faded spruces.

"Chris!" he cried thickly. "Chris, — poor old pard, — don't ye know me? Look, man! Herb is right here — Herb Heal, yer old chum. You're 'heap sick' for sure; but we'll haul you off to a settlement or to our camp, and I'll bring Doc along in two days. He'll " —

But Cross-eyed Chris became past hearing, his flicker of strength had failed; he keeled over, and lay, with his limp legs curled up, faint and speechless, upon the dead evergreens.

"You ain't a-going to die!" gasped Herb defiantly. "I'll be jiggered if you be, jest as I've found you! Say, boys! Cyrus! Neal! rub him a bit, will ye? We ain't got no brandy. I'll build a fire, and warm some coffee."

It was strange work for the hands of the Bostonian, and stranger yet for those of young Farrar, — son of an English merchant-prince, — this straightening and rubbing of a dying half-Indian, a "scum," as Herb called him, drunkard, and thief. Yet there was no flash of hesitation on Farrar's part, as they brought their warm friction to bear upon the chill yellow skin, piebald from dirt and the stains of travel, as if it were the very mission which had brought them to Katahdin.

They had grave thoughts meanwhile that the old mountain was decidedly gloomy in its omens, first a thunder-storm and then a tragedy; for, rub as they might with brotherly hands, they could not pass their own warmth into the body of the half-breed, though he still lived.

But the mountain had not ended its terrors yet.

Its mumbling lips began to speak, with a threatening, low at first like muttered curses, but swelling into a nameless noise — a rumbling, pounding, creeping, crashing.

"Great Governor's Ghost! what's that?" gasped Cyrus, stopping his rubbing. "Pamolah or some other fiend seems to be bombarding us from the top now."

"It's more thunder rolling over us," said Neal; but as he spoke his tongue turned stiff with fear.

"Sounds as if the whole mountain was tumbling to pieces. Perhaps it's the end of the world," suggested Dol, as a succession of booming shocks from above seemed to shake the camping-ground under his feet.

There was one second of awful indecision. The boys looked at each other, at the dying man, at the roof above them, in the stiffness of uncertain terror.

Then a figure leaped into their midst, with an armful of dry sticks, which he dashed from him. It was Herb, with the fuel for a fire.

And, for the first and last time in his history, so far as these friends of his knew it, there was that big fear in his face which is most terrible when it looks out of the eyes of a naturally brave man.

"Boys, where's yer senses?" he yelled cuttingly. "Out, for your lives! Run! There's a slide above us on the mountain!"

"Him?" questioned Cyrus's stiff lips, as he pointed to the breathing wreck on the spruce boughs. "He's not dead yet."

"D'ye think I'd leave him? Clear out of this camp — you, or we'll be buried in less'n two minutes! To the right! Off this ridge! Got yer rifles? I'm coming!"

The woodsman flung out the words while his brawny arms hoisted the body of his old chum. His comrades had already disappeared when he turned and sprang for the camp-door with his limp burden, but his moccasined foot kicked against something.

A great hiccough which was almost a sob rose from Herb's throat. It was his one valuable possession, his 45–90 Winchester rifle, his second self, which he had rested against the log wall.

"Good-by, Old Blazes!" he grunted. "You never went back on me, but I can't lug him and you! My stars! but that was a narrow squeak."

For, as he cleared the camping-ground with a blind dash, with head bent and tongue caught between his clenched teeth, with a boom like a Gatling gun, a great block of granite from the summit of Katahdin struck the rock which sheltered the old camp, breaking a big piece off it, and shot on with mighty impetus down the mountain.

An avalanche of loose earth, stones, and bushes, brought down by this battering-ram of the landslide, piled themselves upon the log hut, smashing to kindling-wood its walls, which had stood many a hard storm, burying them out of sight, and flinging wide showers of dust and small missiles.

A scattered rain of clay caught Herb upon the head, and lodged, some of it, on the little pack containing axe and lunch which was strapped upon his shoulders.

He shook. His grip loosened. The limp, dragging body in his arms sank until the feet touched the earth.

But with the supreme effort, moral and physical, of his life, the forest guide gathered it tight again.

"I'll be blowed if I'll drop him now," he

gasped. "He ain't nothing but a bag o' bones, anyhow."

Only a strong man in the hour of his best strength could have done it. With a defiant snort Herb charged through the choking dust-clouds, pelted by flying pebbles, sods, and fragments of sticks.

"This way, boys!" he roared, after five straining, staggering minutes, as he caught a glimpse of his comrades ahead, tearing off to the right, as he had bidden them. "You may let up now. We're safe enough."

They faced back, and saw him make a few reeling, descending steps, then lay what now seemed to be an out-and-out lifeless man on a bed of moss beneath a dwarfed spruce.

The nerves of the three were in a jumping condition, their brains felt befuddled, and their hearts sinking and melting in the midst of their bones, from the astounding shock and terror of the land-slide. But, as they beheld the guide deposit his burden, with its helplessly trailing head and limbs, a cheer in unsteady tones rang above the slackening rattle of earth and stones, and the far-away boom of the granite-block as it buried itself in the forest beneath.

"Hurrah! for you, Herb, old boy," yelled

Cyrus triumphantly. "That was the grittiest thing I ever saw done! Hurrah! Hurrah! Hoo-ray!"

The English boys, open-throated, swelled the peal.

But their cheering broke off as they came near, and saw the mask-like face over which Herb bent.

"Is he gone, poor fellow?" asked Garst. "What do you suppose caused it — the slide?"

"Why, it was a thundering big lump of granite from the top o' the mountain," answered Herb, replying to the second question. "That plaguy heavy rain must ha' loosened the earth around it the clay and bushes that kep' it in place. So it got kind o' top-heavy, and came slumping and pitching down, slow at first, and then a'most as quick as a cannon-ball, bringing all that pile along with it. I've seen the like before; but, sho! I never came so near being buried by it."

He pointed as he spoke to the late camping-ground, with its lodgment of clay, sods, pygmy trees, and pieces of rock, big and little.

"The old camp's clean wiped out, boys," he said; "and I guess one of the men that

"HERB CHARGED THROUGH THE CHOKING DUST-CLOUDS."

built it is gone, or a'most gone, too. Stick your arm under his head, Cyrus, while I hunt for some water."

Garst did as he was bidden, but his help was not needed long. The guide went off like a racer, covering the ground at a stretching gallop. He remembered well the clear Katahdin spring, which had supplied the home-camp during that long-past trapping winter. He returned with his tin mug full.

When the ice-cold drops touched Chris's forehead, and lay on his parted lips, gem-like drops which he was past swallowing, his malformed eyes slowly opened. There was intelligence in them, shining through the gathering death-film, like a sinking light in a lantern.

He was groping in the dim border-land now, and in it he recognized his old partner with shadowy wonder; for delirium was past, with the other storms of a storm-beaten life.

"Herb," he gurgled in snatches, the words being half heard, half guessed at, "'twas I — took 'em — the skins — an' the antlers. I wanted — to get — to the ole camp — an' let you — take it out o' me — afore I — keeled over."

Herb had taken Cyrus's place, and was upholding him with a tenderness which showed that the guide's heart was in this hour melted to a jelly. Two tears were dammed up inside his eyelids, which were so unused to tears that they held them in. He neither wiped nor winked them away before he answered: —

"Don't you fret about that — poor kid. We'll chuck that old business clean out o' mind. You've jest got to suck this water and try to chipper up, and — we'll make camp together again."

But Herb knew as well as he knew anything that the man who had robbed him was long past "chippering up," and was starting alone to the unseen camping-grounds.

"How long since you got back here?" he asked, close to the dulling ear.

"Couldn't — keep — track — o' days. Got — turned — round — in woods. Lost — trail — heap — long — getting — to — th' old — camp."

The words seemed freezing on the lips which uttered them. Herb asked no more questions. Silence was broken only by the rolling voice of the land-slide, which had not yet ceased. Occasional volleys of loose earth and stones, dislodged or shaken by the down-plunging granite, still kept falling at intervals on the buried camp.

At one unusually loud rattle, Chris's lips moved again. In those strange gutturals which the boys had heard in the hut, he rumbled an Indian sentence, repeating it in English with scared, breaking breaths.

It was a prayer of her tribe which his mother had taught him to say at morning and eve: —

"God — I — am — weak — Pity — me!"

"Heap — noise! Heap — dark!" he gasped. "Can't — find — th' old — camp."

"You're near it now, old chum," said Herb, trying to soothe him. "It's the home-camp."

"We'll — camp — to-ge-ther?"

"We will again, sure."

The last stone pounded down on the heap above the old camp; and Herb gently laid flat the body of the man he had sworn to shoot, closed the malformed eyes, and turned away, that the fellows he was guiding might not see his face.

## CHAPTER XXIII.

### BROTHERS' WORK.

THEY buried Chris upon Katahdin's breast. It was a good cemetery for woodsmen, so Herb said, granite above and forest beneath.

But, good or bad, this was the one thing to be done. An attempt to transfer the body to a distant settlement would be objectless labor; for, as far as the guide knew, the half-breed had not a friend to be interested in his fate, father and mother having died before Herb found him in the snow-heaped forest.

There were three reliable witnesses, besides the man who was known to have a grudge against him, to testify as to the cause and manner of his death when the party returned to Greenville; so no suspicious finger could point at Herb Heal, with a hint that he had carried out his old threat.

How long Chris, in lonely, crazed repentance, had sheltered in the camp on the mountain-side could only be a matter of guess. Herb inclined to think that he had been there for weeks, — months, perhaps, — judging from the withered spruce bed and the dry boughs and sticks upon the camping-ground, which had evidently been gathered and broken for fuel. His ravings made it clear that, on returning to the old haunts after years of absence, he had missed the trail he used to know, and wandered wearily in the dense woods about the foot of Katahdin before he escaped from the prison of trees, and climbed to the hut he sought.

Such wanderings, Herb declared, generally ended in "a man having wheels in his head," being half or wholly insane, though he might keep sufficient wits to provide himself with food and warmth, as Chris had done while his strength held out. This was not long; for the half-breed's words suggested that he felt near to the great change he roughly called "keeling over," when he started to find his cheated partner.

But Cyrus, while he watched the guide making preparations for the mountain burial, pictured the poor weakling tramping for hundreds of miles through rugged forest-land, doubtless with aching knee-joints and feet, that he might make upon his own skin justice for the skins which he had stolen, and so, in the only way he knew, square things with his wronged chum. And the city man thought, with a tear of pity, that even that poor drink-fuddled mind must have been lit by some ray of longing for goodness.

It was a strange funeral.

The guide chose a spot where the earth had been much softened by the recent rain; and, with the ingenuity of a man accustomed to wilderness shifts, he broke up the drenched ground with the axe which he took from his shoulders.

That axe, which had so often made camp, had never before made a grave; the Farrars doubted that it ever would. But Herb worked away upon his knees, moisture dripping from his skin, putting sorrow for years of anger into every blow of his arms. Then, stopping a while, he went off down the mountain to the nearest belt of trees, and cut a limb from one, out of which, with his hunting-knife, he fashioned a rude wooden implement, a cross between a spade and shovel.

With this he scooped out the broken earth until a grave appeared over three feet deep. He lined it with fragrant spruce-boughs from the wind-beaten tangle below.

These Cyrus and Dol had busied themselves in cutting. Neal thought of other work for his fingers. Getting hold of Herb's axe when the owner was not using it, he felled one of the dwarf white birches. Out of its light, delicate wood, with the help of his big pocket-knife and a ball of twine that was hidden somewhere about him, he made a very presentable cross, to point out to future hunters on Katahdin the otherwise unmarked grave.

He was a bit of a genius at wood-carving, and surveyed his work with satisfaction when he considered it finished, having neatly cut upon it the name, "Chris Kemp," with the date, "October 20th, 1891."

"Couldn't you add a text or motto of some kind?" suggested Dol, glancing over his shoulder. "'Twould make it more like the things one sees in cemeteries. You're such a dab at that sort of work."

"Can't think of anything," answered the elder brother.

Then, with a sudden lighting of his face, he seized the knife again, and worked in, in fine lettering, the frightened prayer he had heard on the half-breed's lips: —

"God, I am weak; pity me!"

Herb and Cyrus lowered the body into its resting-place, and covered it with the green spruces.

The four campers knelt bare-headed by the grave.

"Couldn't one of you boys say a bit of a prayer?" asked Herb in a thick voice. "I ain't used to spouting."

All former help had been easily given. This was a harder matter, yet not so difficult as it would have been amid a city congregation.

Garst tried to recall some suitable prayer from a funeral service; so did Neal. Both failed.

But here upon Katahdin's side, where, in the large forces of storm and slide, in forest and granite, through every wind-swept bush, waving blade, and tinted lichen, breathed a whisper from God, it seemed no unnatural thing for a man or a boy to speak to his Father.

"Can't one of you fellers say a prayer?" asked Herb again.

Then the river of feeling in Cyrus broke the dam of reserve, and flowed over his lips in a prayer such as he had never before uttered.

It was the prayer of a son who was for the minute absorbed in his Father.

It left the five, those who were camping here and one who had gone to unseen camping-grounds, with son-like trust to the Father's dealings.

Herb and the Farrars responded to it with heart-eager "Amens!" the fervor of which was new to their lips.

"I thank you as if he were my own brother, boys," said the woodsman, while he filled in the grave, and planted Neal's cross at its head. "Sho! when it comes to a time like we've been through to-day, a man, if he has anything but a gizzard in him, must feel as how we're all brothers,—every man-jack of us,— white men, red men, half-and-half men, whatever we are or wherever we sprung."

"A fellow is always hearing that sort of thing," said Neal Farrar to Cyrus. "But I'm blessed if I ever felt it stick in me before! that we're all of the one stuff, you know—

we and that poor beggar. Some of us seem to get such precious long odds over the others."

"All the more reason why we should do our level best to pull the backward ones up to us," answered the American.

The words struck into the ears of Dol — that youngster listening with a soberness of attention seldom seen in his flash-light eyes.

A few years afterwards, when Neal Farrar was a newly blown lieutenant in his Queen's Twelfth Lancers, as full of heroic impulses and enthusiasms as a modern young officer may be, — while his half-fledged ambitions were hanging on the chances of active service, and the golden, remote possibility of his one day being a V. C., — there was a peaceful honor which clung to him unsought.

During his first year of army life, he became the paragon of every poor private and raw recruit struggling with the miseries of goose-step, with whom he came even into momentary contact. For sometimes through a word or act, sometimes through a flash of the eye, or a look about the mouth, during the brief interchange of a military salute, these "backward ones" saw that the progressive young officer looked on them, not as men-machines, but as brothers, as important in the great schemes of the nation and the world as he was himself; that he was proud to serve with them, and would be prouder still to help them if he could.

It was an understanding which inspired many a tempted or newly joined fellow to drill himself morally as his sergeant drilled him physically, with a determination to become as fine a soldier and forward a man as his paragon.

But only one American friend of Lieutenant Farrar's, who has let out the secret to the writer, knows that the binding truth of human brotherhood was first born into him when, on Katahdin's side, he helped to bury a thieving half-Indian.

## CHAPTER XXIV.

### " KEEPING THINGS EVEN."

"NOW, you mustn't be moping, boys, because of this day's work that you took a hand in, and that wasn't in your play-bill when you come to these woods. We'll have to try and even things up to-morrow with some big sport. You look kind o' wilted."

So said Herb when the tired party were half-way back to camp, doing the descent of the mountain in a silence clouded by the scene which they had been through.

The woodsman seemed troubled with a rasping in his throat. He cleared it twice and spat before he could open a passage for a decently cheerful voice in which to suggest a rise of spirits. But Herb was too faithful a guide to bear the thought that his employers' trip should end in any gloom because the one painful chapter in his own life had closed forever. Moreover, although more than once, as he fought his way through a jungle or jumped a windfall, something nipped his heart, pinching him up inside, and making his eyes leak, he felt that the thing had ended well for him — and for Chris.

Herb, in his simple faith, scarcely doubted that the old chum, whom he had forgiven, had reached a Home-Camp where his broken will and stunted life might be repaired, and grow as they had poor chance to grow here.

"Say, boys!" he burst forth, a few minutes after his protest against "moping," and when the band were within sight of the spring whence they had started, an age back, as it seemed, on the trail of the moose. "Say, boys! I've been all these years raging at Chris. Seems to me now as if he was a poor sort of overgrowed baby, and not so bad a thief as the chump who gave him that whiskey, and stole his senses. It's a thundering big pity that man hadn't the burying of him to-day.

"He was always the under dog, — was

Chris," he went on slowly, as if he was seeking from his own heart an excuse for those unforeseen impulses which had worked it and his body during the past five hours. "Whites and Injuns jumped on him. They said he was criss-cross all through, same as his eyes. But he warn't. Never seed a half-breed that had less gall and more grit, except when the hanker for whiskey would creep up in him, and boss him. He could no more stand agen it, and the things it made him do, than a jack-rabbit."

"Another reason why we Americans ought to feel our responsibility towards every man in whose veins runs Indian blood, a thousand times more hotly than we do!" burst out Cyrus. "It maddens a fellow to think that we made them the under dogs, and as much by giving them a 'boss,' as you say, in fire-water, as by anything else."

"I kind o' think that way myself sometimes," said Herb.

And there was silence until the guide cried : —

"Here's our camp, boys. I'll bet you're glad to see it. I must get the kettle, and cruise off for water. 'Tain't likely I'll trust one of you fellers after last night. But you can hustle round and build the camp-fire while I'm gone."

Herb had a shrewd motive in this. He knew that there is nothing which will cure the blues in a camper, if he is touched by that affliction, rare in forest life, like the building of his fire, watching the little flames creep from the dull, dead wood, to roar and soar aloft in gold-red pennons of good cheer.

The result proved his wisdom. When he returned in a very short time from that ever-to-be-famous spring, with his brimming kettle, he found a glorious fire, and three tired but cheerful fellows watching it, its reflection playing like a jack-o'-lantern in each pair of eyes.

"Now I'll have supper ready in a jiffy," he said. "I guess you boys feel like eating one another. Jerusha! we never touched our snack — nary a crumb of it."

In the strange happenings and chaotic feelings of the day, hunger, together with the bread and pork for satisfying it which Herb had carried up the mountain, were forgotten until now.

"Never mind! We'll make up for it. Only hurry up!" pleaded Dol. "We're like bears, we're so hungry."

"Like bears! You're a sight more like calves with their mouths open, waiting for something to swallow," answered Herb, his eyes flashing impudence, while, with an energy apparently no less brisk than when he started out in the morning, he rushed his preparations for supper.

"Say I'm like a Sukey, and I'll go for you!" roared Dol, a gurgling laugh breaking from him, the first which had been heard since the four struggled through that tangle on Katahdin to a sight of the old camp.

Once or twice during supper the mirth, which had been frozen in each camper's breast by a sight of the drifted wreck of a human life, warmed again spasmodically. Herb did his manly best to fan its flame, though his heart was still pinched by a feeling of double loss.

Later in the evening, when the party were huddling close to the camp-fire, he lifted his right hand and looked at it blankly.

"My!" he gasped, "but it will feel awful queer and empty without Old Blazes. That rifle was a reg'lar corker, boys. I was saving up for three years to buy it. An' it never went back on me. Times when I've gone far off hunting, and had nary a chance to speak to a human for weeks, I'd get to talking to it like as if 'twas a living thing. When I wasn't afeard of scaring game, I'd fire a round to make it answer back and drive away lonesomeness. Folks might ha' thought I was loony, only there was none to see. Well, it's smashed to chips now, 'long with the old camp."

"What awfully selfish jackasses we were, to skip off with our own rifles, and never think of yours, or that you couldn't save it, carrying that poor fellow! I feel like kicking myself," said Cyrus, sharp vexation in his voice. "But that slide business sprang on us so quickly. The sudden rumbling, rattling, and pounding jumbled a fellow's wits. I scarcely understood what was up, even when we were scooting for our lives."

"I felt a bit white-livered myself, I tell ye; and I'm more hardened to slides than you are," was the woodsman's answer.

The confession, taken in the light of his conduct, made him doubly a hero to his city friends.

They thought of him staggering along the mountain, blinded, bewildered, pelted by clay, with that dragging burden in his arms, a heart tossed by danger's keenest realization in his breast. And they were silent before the high courage which can recognize fear, yet refuse to it the mastery.

Neal, whose secret musings were generally crossed by a military thread, seeing that he had chosen the career of a cavalry-soldier, and hoped soon to enter Sandhurst College, stared into the heart of the camp-fire, glowering at fate, because she had not ordained that Herb should serve the queen with him, and wear upon his resolute heart — as it might reasonably be expected he would — the Victoria Cross.

Young Farrar's feeling was so strong that it swept his lips at last.

"Blow it all! Herb," he cried. "It's a tearing pity that you can't come into the English Lancers with me. I don't suppose I'll ever be a V. C., but you would sooner or later as sure as gun's iron."

"A 'V. C.!' What's that?" asked Herb.

"A Vigorous Christian, to be sure!" put in Cyrus, who was progressive and peaceful, teasingly.

But the English boy, full of the dignity of the subject to him, summoned his best eloquence to describe to the American backwoodsman that little cross of iron, Victoria's guerdon, which entitles its possessor to write those two notable letters after his name, and which only hero-hearts may wear.

But a vision of himself, stripped of "sweater" and moccasins, in cavalry rig, becrossed and beribboned, serving under another flag than the Stars and Stripes, was too much for Herb's gravity and for the grim regrets which wrung him to-night.

"Oh, sugar!" he gasped; and his laughter was like a rocket shooting up from his mighty throat, and exploding in a hundred sparkles of merriment.

He laughed long. He laughed insistently. His comrades were won to join in.

When the fun had subsided, Garst said:—

"Herb Heal, old man, there's something in you to-night which reminds me of a line I'm rather stuck on."

"Let's have it!" cried Herb.

And Cyrus quoted: —

> "As for this here earth,
> It takes lots of laffin' to keep things even!"

"Now you've hit it! The man that wrote that had a pile o' sense. Come, boys, it's been an awful full day. Let's turn in!"

As he spoke, Herb began to replenish the

fire, and make things snug in the camp for the night.

But shortly after, when he threw himself on the spruce-boughs near them, the boys heard him murmur, deep in his throat, as if he took strength from the words:—

"It takes lots of laffin' to keep things even!"

## CHAPTER XXV.

#### A LITTLE CARIBOU QUARREL.

BUT things on this old planet seemed even enough the next day, when, after a dozen hours of much needed sleep, the campers' eyes opened upon a scene which might have stirred any sluggish blood — and they were not sluggards.

A fresh breath of frost was in the air to quicken circulation and hunger. Under a smiling sun an October breeze frolicked through leaves with tints of fire and gold, humming, while it swiftly skimmed over their beauties, as if it was reading a wind's poem of autumn.

Katahdin looked as though it had suddenly taken on the white crown of age, with age's stately calm. The weather had grown colder during the night. Summer — the balmy Indian summer, with its late spells of sultriness — had taken a weeping departure yesterday. To-day there was no threatening of rain-storm or slide. The mountain's principal peaks had fleecy wraps of snow.

"Ha! Old Katahdin has put on its nightcap," exclaimed Cyrus, when the trio issued from their tent in the morning. "Listen, you fellows! This is the 21st of October. I propose that we start back to our home-camp to-morrow. It will take us two days to reach Millinokett Lake. Then we'll set our faces towards civilization the first week in November, or thereabouts."

"Oh, bother it! So soon!" protested Dol.

"Now, Young Rattlebrain,"—Garst took the calm tone of leadership, — "please consider that this is the first time you've camped out in Maine woods. You might find it fun to be snowed up in camp during a first fall, and to tramp homewards through a thawing slush. But your father wouldn't relish its effects on your British constitution. And out here — once we're well into November — there's no knowing when the temperature may drop to zero with mighty short notice. I've often turned in at night, feeling as if I were on 'India's coral strands' and woke up next morning thinking I had popped off in my sleep to 'Greenland's icy mountains.' Herb Heal! you know what tricks a thermometer, if we had one, might play in our camp from this out; talk sense to these fellows."

Herb, who had risen an hour before his charges, had already fetched fresh water, coaxed up the fire, and was busily mixing flapjacks for breakfast. His ears, however, had caught the drift of the talk.

"Guess Cyrus is right," he said. "Seeing as it's the first time you Britishers have slept off your spring mattresses, I'd say, light out for the city and steam-heat afore the snow comes. Oh! you needn't get your mad up. I ain't thinking you'd growl at being snowed in. I know better.

"By the great horn spoon! I b'lieve I'll go right along to Greenville with you," exclaimed the guide a minute later. "I might get a chance to pick up a bargain of a second-hand rifle there. And I guess you'd be mighty sick o' your luck, Dol, if you had to lug them moose-antlers part o' the way yer-self. I ain't stuck on carrying 'em either, if we can get a jumper."

But there was a third reason, still more powerful than these two, why he should make a trip to the distant town, which stirred Herb's mind while he stirred his cakes. His sturdy sense told him that it would be well he should put in an appearance when Cyrus made a statement before the Greenville coroner as to the cause and manner of Chris's death.

"Now, you boys, we don't want no fooling this blessed day," he said, when breakfast was in order, and the campers were emptying for the second time their tin mugs of coffee. "There's sport before us — tearing good sport. Whatever do you s'pose I come on this morning when I was cruising over the bog for water? Caribou-tracks! Caribou-tracks, as sure as there's a caribou in Maine!

"Who's for following 'em? We hain't got much provisions left; and I guess a chunk of broiled caribou-steak about as big as a horse's upper lip would cheer each of us up, and make us feel first-rate. What say, boys?"

"By all that's glorious!" ejaculated Cyrus, his eyes striking light. "Caribou-signs! Of course we'll follow them. A bit of fresh meat would be pretty acceptable, and a good view of a herd of caribou would be still more so—to me, at any rate. That would just about top off our exploring to a T."

"We've got to be mighty spry, then," said the woodsman, lurching to his feet, muscles swelling, and nostrils spreading like a sleuth-hound's. "If you want caribou, you've got to take 'em while they're around. Old hunters have a saying: 'They're here to-day, to-morrow nowhere.' And that's about the size of it."

"Let's start off this minute!" Dol jerked out the words while he bolted the last salt shreds of his pork. "Hurry up, you fellows! You're as slow as snails. I'd eat the jolliest meal that was ever cooked in three minutes."

"No wonder you squirm and shout all night, then, until sane people with good digestions feel ready to blow your head off," laughed Cyrus, who was one of the laggards; but he disposed of the last mouthfuls of his own meal with little regard for his digestive canal.

In rather less than twenty minutes the four were scanning with wide eyes certain fresh foot-marks, plainly printed on a patch of soft oozing clay, midway on the boggy tract.

"Whew! Bless me! Those caribou-tracks?" Cyrus caught his breath with amazement while he crouched to examine them. "Why, they're bigger than any moose-tracks we've seen!"

"Isn't that great?" gasped Dol.

"Well, come to think of it, it is," answered the guide, in the stealthy tones of an expectant hunter; "for a full-grown bull-caribou don't stand so high as a full-sized moose by two or three feet, and he don't weigh more'n half as much. Still, for all that, caribou deer beat every other animal of the deer tribe, so far's I know, in the size of their hoofs, as you'll see bime-by if luck's with us! And my stars! how they scud along on them big hoofs. I'd back 'em in a race against the smartest of your city chaps that ever spun through Maine on his new-fangled 'wheel,' that he's so sot on."

Garst, who was an enthusiastic cyclist, with a gurgle of unbelieving mirth, prepared to dispute this. There might have ensued a wordy sparring about caribou versus bicycle, had not the guide been impressed with the necessity for prompt action at the expense of speech.

"We must quit our talk and get a move

on," he whispered, and led the forward march across the bog, his eyes every now and again narrowing into two gleaming slits, as if he were debating within himself, while he studied the ground or some bush which showed signs of being nibbled or trampled. Then he would sweep the horizon with long-range vision.

But not a tuft of hair or glancing horn hove in sight.

The marsh was left behind. The hoof-marks were lost in a wide meadowy sweep of open ground, bounded at a distance by an irregular line of hills, sparsely covered with spruce-trees.

Towards these Herb headed, leaving Katahdin away back in the rear.

"'Shaw! I'm afeard they're 'nowhere' by this time," he whispered, when the hunters reached the rising ground, glancing at Dol, who stepped lightly beside him.

The boy's lips parted to breathe out compressed disappointment; but his answer was lost in a sharp whirr! whirr! and a sudden flutter of wings above his head. His eyes went aloft towards a bough about eight feet from the ground. So did Herb's, and lit with a new, whimsical hope.

"A spruce partridge!" hissed the guide, his voice thrilling even in its stealthy whisper. "That's luck — dead sure! The Injuns say, 'The red eye never tells a lie;'" and the woodsman pointed out the strip of bare red skin above the beady eyes of the bird, which cuddled itself on its branch, and looked down at them unfrighted.

Dol Farrar, who in this region of moose-birds and moose-calls could believe in anything, felt both his spirits and credulity rise together. He managed to keep abreast of the trained hunter, as the latter, with swift, stretching, silent steps climbed the hill. And he heard the hunter's sudden cluck of triumph as he reached the top, and looked down upon the valley at the other side, the inarticulate sound being followed by one softly rung word, —

"Caribou!"

"Caribou? They look awfully like quiet Alderney cows, except for the big antlers!" The amazed exclamation stirred the English boy's tongue, but he did not make it audible.

Following Herb's example, he stretched himself flat upon his stomach under a spruce, and stared over the brow of the hill at a forest pantomime which was being acted in the valley.

Cautiously slipping from tree to tree, Cyrus and Neal, who had lagged a few steps behind, joined the leaders, and lay low, eagerly gazing too.

On its farther side the hill was yet more sparsely covered, the scattered spruces showing gaps between them where the lumberman's axe had made havoc. Through these openings, which were as shafts of light amid the evergreen's waving play, the hunters saw the sun silver a brown pool in the valley. A few maples and birches waved their shrivelling splendors of scarlet and buff at irregular distances from the water. And in and out among these trees moved in graceful woodland frolic four or five large animals, — perhaps more, — their doings being plainly seen by the watchers on the hill.

Their coats, like those of the smaller deer, were of a brown which seemed to have caught its dye from the autumnal tints surrounding them. In shape they justified Dol's criticism; for they certainly were not unlike cows of the Alderney breed, save for the widely branching horns.

Of the strength of these antlers the hidden

spectators got sudden, startling proof, as the two largest caribou drew off from the rest, and charged each other in a real or sham fight, the battle-clang of their meeting horns sounding far away to the hill-top.

"Them two bulls are having a big time of it. Look at 'em now, with the small one. That's a stranger in the herd," hummed Herb into the ear of the boy next to him, his voice so light and even that it might have been but the murmur of a falling leaf. "It's an all-fired pity that we're jest too far off for a shot."

The "stranger," which the woodsman's long-range eye had singled out, was of a smaller size and paler color than the other caribou; and Herb — who could interpret the forest pantomime far better than he would have explained the acting of human beings on a stage — told his companions in whispers and signs that it was in distressed dread of its company.

The attentions which the rest paid to it seemed at first only friendly and facetious. The two big bulls, after trying their mettle against each other for a minute, separated, and moved towards it, prodded it lightly with their horns, and playfully bit its sides, a sport in which the other members of the herd joined.

"They're playing it, like a cat with a mouse; but I guess they'll murder it in the long run if it's sickly or weak. Caribou are the biggest bullies in these woods — to each other," whispered Herb.

"By the great horn spoon! they're doing for it now," he gasped, a minute later. "Sho! ... if I only had my old Winchester here, I'd soon stop their lynching. Try it, you, Cyrus! You're a sure shot, an' you can creep within a hundred yards of 'em without being scented. Try it, man!"

The guide's flashing eyes and quick signs conveyed half his meaning; his excited sentences were so low that Garst only caught fag-ends of them. But they were emphasized unexpectedly by a faint bleating sound rising from the valley, — the helpless bleat of a buffeted creature.

"We want meat, and I'm going to spring a surprise on those bullies," muttered Cyrus, setting his teeth.

Still lying flat, he shot his eyes down the hill-slope, forming a plan of descent; then he lifted the rifle beside him, and jammed some fresh cartridges into the magazine.

Ere a dozen long breaths had been drawn, he was stealthily moving towards the valley, slipping from spruce to spruce — an arrow-like, unnoticeable figure in his dark gray tweeds.

He was close to the foot of the hill when the three breathless fellows above saw him raise his rifle, just as the unfortunate little caribou, after many efforts to escape, had been beaten to its knees.

"He'll drop one, sure! He's a crack shot — is Cyrus! There! he's drawing bead. Bravo! . . . he's floored the biggest!"

Herb's gusty breath blew the sentences through his nostrils, while the sudden, explosive bang of the Winchester cut through all other sounds, and set the air a-quiver.

Twice Cyrus fired.

The largest bull-caribou leaped three feet upward, wheeled about, staggered to his knees. A third shot stopped his bullying forever.

"Hurrah! I guess you've got the leader — the best of the herd. That other bull was a buster too! You might ha' dropped him, if you'd been in the humor!" bellowed the guide, springing to his legs, and letting out his pent-up wind in a full-blast roar of triumph.

He well knew that Cyrus, " being a queer specimen sportsman," and the right sort after all, would be satisfied with the one inevitable deed of death.

As their leader fell, the caribou raised their heads, stared in stiffened wonder for a few seconds, offering a steady mark for the smoking rifle if it had been in the grasp of a butcher. Then, as though propelled by one shock, they cut for the wood at dazzling speed.

A minute — and they were in the distance as tufts of hair blown before a storm-wind.

The half-killed weakling sought shelter more slowly in another direction.

" Well done, Cy ! "

" Congratulations, old man ! "

" You've got a trophy now. You'll never leave this splendid head behind. My eye, what antlers ! "

Such were the exclamations blown to Garst's ears by the hot breath of his English friends, as they reached his side, and stooped with him to examine the fallen forest beauty.

" No ; I guess we can manage to haul the head back to camp, with as much meat as we need. You'll have your 'chunk of caribou-steak as big as a horse's upper lip,' to-night, Herb, and bigger if you want it. I'm tickled at getting the antlers, especially as I didn't shoot this beauty for the sake of them. I'll hook them on my shoulders when we start back to Millinokett to-morrow."

So answered the successful hunter, tingling with some pride in the skill which, because of his reverence for all life, he generally kept out of sight.

And he stuck to his purpose about the antlers.

Cheered and invigorated by a sumptuous supper and breakfast of broiled caribou-steaks, supplemented by Herb's lightest cakes, and carrying some of the meat with them as provision for the way, the campers accomplished their backward tramp to the log camp on Millinokett Lake in fulness of strength and spirits.

Once or twice during the journey, when the guide was stalking ahead, and thought himself unnoticed, the city fellows saw him lift his right hand and look at it for a full minute. Then it swung heavily back to his side.

"He's missing his rifle, the partner that never went back on him," said Cyrus. "Say, boys! I've got an idea!"

"Out with it if it's worth anything," grunted Dol. "I never have ideas these days. Too much doing. I don't feel as if there was a steady peg in me to hang one on."

"Oh! quit your nonsense, Chick, and listen. Herb will wait for us in a few minutes," was the Boston man's impatient rejoinder.

Then followed a low-toned consultation, in the course of which such talk as this was heard: —

"Our Pater will want to shell out when he hears about Chris."

"So will mine. He'll be for sending Herb a cool five hundred or thousand dollars, right away. And, as likely as not, Herb would feel flaring mad, and ready to chuck it in his face. He's not the sort of fellow to stand being paid by an outsider for a plucky act, done in the best hour of his life."

"Oh, I say! wouldn't it be decenter to manage the thing ourselves, without letting anybody who doesn't know him meddle in it?" This suggestion was in Dol's voice. "Neal and I could draw our allowances for three months in advance; the Pater will be willing enough. We'll be precious hard up without them, but we'll rub through somehow. Then you can chip in an even third,

Cy, and we'll order an A 1 rifle, — the best ever invented, from the best company in America, — silver plate, with his name, — and all the rest of it. I'd swamp my allowance for a year to see Herb's face when he gets it."

"That's the plan! You do have occasional moments of wisdom, Dol; I'll say that much for you," commented the leader. "Well, Herb has taken a special sort of liking to you. You may tip him a hint to wait in Greenville for a few days, and not to go looking for second-hand rifles till he hears from us. Better not say anything until we're just parting. Ten to one, though, you'll blurt the whole thing out in some harebrained minute, or give it away in your sleep."

"Blow me if I do!" answered Dol solemnly.

## CHAPTER XXVI.

### DOC AGAIN.

HERB, turning back at that minute to wait for his party, experienced a shock of curiosity which was new to him, at seeing the three in close counsel, shouldering each other upon a trail a couple of feet wide.

But the sensation passed. Dol for once was not guilty of an indiscretion, waking or sleeping. The woodsman got no hint of what matter had been discussed until more than two weeks later, when he stood in the main street of Greenville, beside a tanned, muscular, newly shaven trio, waiting for their departure for Boston.

A few pleasant days, marked by no particular excitements, had been spent at the log camp on Millinokett after that wonderful trip into the forests of Katahdin. Then the weather turned suddenly blustering and cold; and Cyrus, as captain, ordered an immediate forced march to Greenville.

Under Herb's guidance that march was made with singularly few hardships. He managed to hire a "jumper" from a new settler who had a farm a couple of miles from their camp. This contrivance was a rough sort of sled, formed of two stout ash saplings, and hitched to a courageous horse. The "jumper's" one merit was that it could travel along many a rough trail where wheels would be splintered at the outset. But since, as Herb said, it went at " a succession of dead jumps," no camper was willing to trust his bones to its tender mercies. However, it answered admirably for carrying the tent, knapsacks, and trophies of the party, tightly strapped in place, including Neal's bear-skin, which was duly called for, and the moose-antlers, more precious in Dol's sight than if they had been made of beaten gold.

Thus the campers journeyed homeward with their backs as light as their spirits, caring little for the chills of a couple of nights spent under canvas and rubber coverings.

Two gala evenings they had, — one with Uncle Eb in his bark hut near Squaw Pond, where they were regaled with a sumptuous supper, for "coons war in eatin' order now;" and the second with Doctor Phil Buck at his little frame house near Moosehead Lake.

Dear old Doc was as ever a power, — a power to welcome, uplift, entertain.

The campers sought him immediately on their arrival at Greenville; and he stood by them while Cyrus made a full statement before the local coroner about the death and burial of the half-breed, Chris Kemp, the Farrars and Herb confirming what was said with due dignity.

But dignity was blown to the four winds by the very unprofessional and very woodsman-like cheer that Doc raised, and that was echoed thunderously by Joe Flint and a few other guides and loungers who had collected to hear the story, when Cyrus described the splendid rush which Herb made, with the dying man in his arms, and the clay of the landslide half smothering him.

"I'm sorry I wasn't near to try and do something for the poor fellow," said the doctor, later on, when his friends were gathered round a blazing wood-fire in his own snug house. "But I doubt if I could have helped him. I guess he was born with the hankering for whiskey, and when that is in the mongrel blood of a half-breed it is pretty sure to wreck him some time. We must leave him to God, boys, and to changes larger than we know."

"I've a letter for you, Neal," added the host presently in a lighter tone. "It was directed to my care. It is from Philadelphia, from Royal Sinclair, I think."

Neal slit the envelope which was handed to him, and read the few lines it contained aloud, with a longing burst of laughter.

Royal was as short with his pen as he was dash-away with his tongue. The letter was a brief but pressing invitation to Cyrus and the Farrars to visit their camping acquaintances of the Maine wilds at the Sinclairs' home in Philadelphia before the English boys re-crossed the Atlantic.

"Come you must!" wrote Roy. "We've promised to give a big spread, and invite all the crowd we train with to meet you. We'll have a great old time, and bring out our best yarns. Don't let me catch you refusing!"

"We won't if we can help it," commented Neal; "if only we can coax the Pater to give us another week in jolly America."

Greenville,—"Farewell to the Woods."

The campers slept upon mattresses that night for the first time in many weeks.

The following morning saw them grouped in the main street of Greenville, with Doc and Herb on hand for a final farewell, waiting for the departure of the coach which was to bear them a little part of the way towards Boston civilization.

Dol was turning over in his jostled thoughts the delicate wording of the hint which he was to convey to Herb about the rifle, when he became aware that Doctor Phil was pinching his shoulder, and saying, while he drew Neal's attention in the same way: —

"Well, you fellows! I'm glad to have known you. If you ever come to Maine again, remember that there's one old forest fogy who'll have a delightful welcome for you in his house or camp, not to speak of the thing he calls his heart. And I hope you'll keep a pleasant corner in your memories for our Pine Tree State, and for American States generally, so far as you've seen them."

Dol tried to answer; but recalling the evening when, wrecked at heart, with stinging feet, he had stumbled at last into the trail to Doc's camp, he could only mutter, "Dash it all!" and rub his leaking eyes.

"Of course I'll think in an hour from now of all the things I want to say," began Neal helplessly, and stopped. "But I'll tell you how I feel, Doc," he added, with a sudden rush of breath: "I think I can never see your Stars and Stripes again without taking off my hat to them, and feeling that they're about equal to my own flag."

"Neatly put, Neal! I couldn't have done it better," laughed Cyrus.

"Shake!" and Doc offered his hand in a heart-grip, while the hairs on it bristled. "Boy! long life to that feeling. You men who are now being hatched will show us one day what Young England and Young America, as a grand brotherhood under comrade flags, can do to give this old earth a lift which she has never had yet towards peace and prosperity. We're looking to you for it!"

"Hur-r-r-rup!" cheered Herb, subduing his shout to the requirements of a settlement, but sending his battered hat some ten feet into the air, and recovering it with a dexterous shoot of his long arm, by way of giving his friends an inspiring send-off.

"Tell you what it is!" he said suddenly, turning upon the Farrars, "I never guided

Britishers till now; but, wherever you sprung from, you're clean grit. If a man is that, it don't matter a whistle to me what country riz him."

A few minutes afterwards, with a jingle, jangle, lurch, and rattle, the stage-coach was swaying its way out of Greenville. Dol, stooping from his seat upon it, gripped the guide's hand in a wringing good-by.

"Herb," he said, "we three fellows want you to stay here for a few days, and not to do anything about a second-hand rifle until you hear from us. Mind!"

And so it happened that, ten days or so later, while the three were enjoying the hospitalities of the Sinclairs and "their crowd" in the Quaker City, Herb, who was still in Greenville, waiting for a fresh engagement as guide, was accosted by the driver of the coach from Bangor.

"Herb Heal, here's a bully parcel for you," said the Jehu, with a knowing grin. "Came from Boston, I guess. I war booked to take pertik'lar care of it."

And Herb, feeling his strong fingers tingle, undid many wrappers, and hauled out, before the eyes of Greenville loungers, a rifle such as it is the desire of every Maine woodsman's heart to possess.

A best grade, 45–90, half-magazine Winchester it was, fitted with shot-gun stock and Lyman sights, and bearing a gleaming silver plate, on which was prettily lettered: —

HERB HEAL.<br>
IN MEMORY OF OCTOBER, 1891.

Underneath was engraved a miniature pine, its trunk bearing three sets of initials.

Herb stalked straight off a distance of one mile to Doctor Buck's house, pushed the door open as if it had been the door of a wilderness camp, and shot himself into Doc's little study.

"Look what those three gamy fellows have sent me," he said; and his eyes were now like Millinokett Lake under a full sun-burst. "I thought the old one was a corker, but this" —

Here the woodsman's dictionary gave out.

## CHAPTER XXVII.

### CHRISTMAS ON THE OTHER SIDE.

"'CHRISTMAS, 1893.' Those last two figures are a bit crooked; aren't they, Dol?" said a tall, soldierly fellow, who was no longer a boy, yet could scarcely in his own country call himself a man.

He read the date critically, having fixed it as the centre-piece in a festive arch of holly and bunting, which spanned the hall of a mansion in Victoria Park, Manchester.

"I believe that's better," he added, straightening a tipsy "93," and bounding from a chair-back on which he was perched, to step quickly backward, with a something in gait and bearing that suggested a cavalry swing.

"'Christmas, 1893,'" he read musingly

again. "Goodness! to think it's two years since we laid eyes on old Cyrus, and that he has landed on English soil before this, may be here any minute — and Sinclair too. I guess"— these two words were brought out with a smile, as if the speaker was putting himself in touch with the happiness of a bygone time—"I guess that 'Star-Spangled Banner' will look home-like to them."

And Neal Farrar, just back for a short vacation from Sandhurst Military College, twice gravely saluted the gay bunting with which his Christmas arch was draped, where the Union Jack of old England kissed the American Stars and Stripes.

"I say!" he exclaimed, turning to a tall youth, who had been inspecting his operations, "that Liverpool train must be beastly late, Dol. Those fellows ought to be here before this. The Mater will be in a stew. She ordered dinner at five, as the youngsters dine with us, of course, to-day, and it's past that now."

"Hush! will you? I'll vow that cab is stopping! Yes! By all that's splendid, there they are!" and Dol Farrar's joy-whoop rang through the English oaken hall with scarcely less vehemence than it had rung in former days through the dim aisles of the Maine forests.

A sound of spinning cab-wheels abruptly stopping, a noise of men's feet on the steps outside, and the hall-door was flung wide by two pairs of welcoming hands.

"Cyrus! Royal! Got here at last? Oh! but this is jolly."

"Neal, dear old boy, how goes it? Dol, you're a giant. I wouldn't have known you."

Such were the most coherent of the greetings which followed, as two visitors, in travelling rig, their faces reddened by eight days at sea in midwinter, crossed the threshold.

There could be no difficulty in recognizing Cyrus Garst's well-knit figure and speculative eyes, though a sprouting beard changed somewhat the lower part of his face. And if Royal Sinclair's tall shoulders and brand-new mustache were at all unfamiliar, anybody who had once heard the click and hum of his hasty tongue would scarcely question his identity.

The Americans had steamed over the Atlantic amid bluster of elements, purposing a tour through southern France and Italy. And they were to take part, before proceeding to the Continent, in the festivities of an English Christmas at the Farrars' home in Manchester.

"Oh, but this is jolly!" cried Neal again, his voice so thickened by the joy of welcome that — embryo cavalry man though he was — he could bring out nothing more forceful than the one boyish exclamation.

Dol's throat was freer. Sinclair and he raised a regular tornado in the handsome hall. Questions and answers, only half distinguishable, blew between them, with explosions of laughter, and a thunder of claps on each other's shoulders. When their gale was at its noisiest, Royal's part of it abruptly sank to a dead calm, stopped by "an angel unawares."

A girl of sixteen, with hair like the brown and gold of a pheasant's breast, opened a drawing-room door, stepped to Neal's side, and whispered, —

"Introduce me!"

"My sister," said Neal, recovering self-possession. "Myrtle, I believe I'll let you guess for yourself which is Garst and which is Sinclair."

"Well, I've heard so much about you for the past two years that I know you already, all but your looks. So I'm sure to guess right," said Myrtle Farrar, scrutinizing the Americans with a pretty welcoming glance, then giving to each a glad hand-shake.

Royal's tongue grew for once less active than his eyes, which were so caught by the golden shades on the pheasant-like head that for a minute he could see nothing else. Even Cyrus, who was accustomed to look upon himself as the cool-blooded senior among his band of intimates, tingled a little.

"You're just in time for dinner — I'm so glad," laughed Miss Myrtle. "A Christmas dinner with a whole tribe of Farrars, big and little."

"But our baggage hasn't come on yet," answered Garst ruefully. "Will Mrs. Farrar excuse our appearing in travelling rig?"

"Indeed she will!" answered for herself a fair, motherly-looking English woman, as pretty as Myrtle save for the gold-brown hair, while she came a few steps into the hall to welcome her sons' friends.

Five minutes afterwards the Americans found themselves seated at a table garlanded with red-berried holly, trailing ivy, and pearl-eyed mistletoe, and surrounded by a round dozen of Farrars, including several youngsters whose general place was in school-room or nursery, but who, even to a tot of three, were promoted to dine in splendor on Christmas Day.

"Well, this is festive!" remarked Cyrus to Myrtle, who sat next to him, when, after much preparatory feasting, an English plum-pudding, wreathed, decorated, and steaming, came upon the scene. Fluttering amid the almonds which studded its top were two wee pink-stemmed flags. And here again, in compliment to the newly arrived guests, the "Star-Spangled Banner" kissed the English Union Jack.

"Say, Neal!" exclaimed Cyrus, his eyes keenly bright as he looked at the toy standards, "wouldn't this sort of thing delight our friend Doc? By the way, that reminds me, I have a package for you from him, and a message from Herb Heal too. Herb wants to know 'when those gamy Britishers are coming out to hunt moose again?' And Doc has sent you a little bundle of beaver-clippings. They are from an ash-tree two feet in circumference, felled by that beaver colony which we came across near the *brûlée* where you shot your bear and covered yourself with glory. Doc asked you to put the wood in sight on Christmas Night, and to think of the Maine woods."

"Think of them!" Neal ejaculated. "Bless the dear old brick! does he think we could ever forget them and the stunning times we had in camp and on trail?"

www.ingramcontent.com/pod-product-compliance
Lightning Source LLC
Chambersburg PA
CBHW032030220426
43664CB00006B/428